METALLIC NANOSTRUCTURES

METALLIC NANOSTRUCTURES

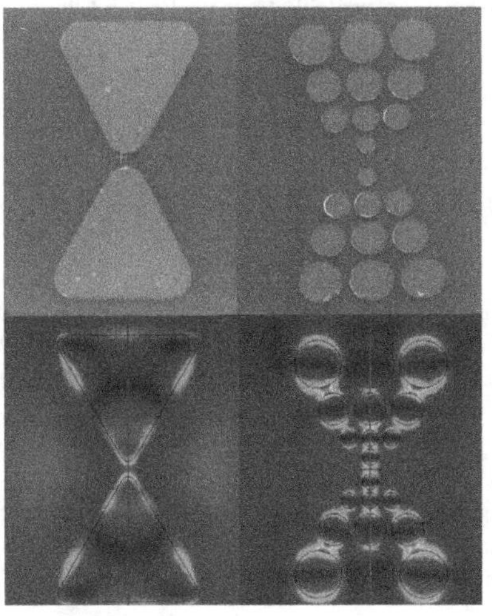

Francisco Javier Gonzalez

University of Central Florida, USA

World Scientific

NEW JERSEY · LONDON · SINGAPORE · BEIJING · SHANGHAI · TAIPEI · CHENNAI

Published by

World Scientific Publishing Co. Pte. Ltd.

5 Toh Tuck Link, Singapore 596224

USA office: 27 Warren Street, Suite 401-402, Hackensack, NJ 07601

UK office: 57 Shelton Street, Covent Garden, London WC2H 9HE

Library of Congress Control Number: 2025932978

British Library Cataloguing-in-Publication Data
A catalogue record for this book is available from the British Library.

ISBN 978-981-98-1177-9 (hardcover)
ISBN 978-981-98-1178-6 (ebook for institutions)
ISBN 978-981-98-1179-3 (ebook for individuals)

For any available supplementary material, please visit
https://www.worldscientific.com/worldscibooks/10.1142/14272#t=suppl

Desk Editor: Muhammad Ihsan Putra

Typeset by Stallion Press
Email: enquiries@stallionpress.com

Contents

Part 1

Introduction

In this part, we embark on a journey through the foundational concepts of nanoscience and nanotechnology, exploring how the relentless drive for miniaturization has transformed our understanding of materials and their applications. We will delve into the prophetic insights of Richard Feynman, whose iconic 1959 lecture at Caltech, "There's Plenty of Room at the Bottom," ignited a revolution in how we think about manipulating matter at the atomic level. In this section, we will also examine the implications of Moore's law, which predicts the exponential growth of computing power through smaller and smaller components, and we will uncover the unique size-dependent properties that emerge in nanomaterials. These properties not only enhance functionality but also open new avenues for innovation across various fields. Intriguingly, metallic nanoparticles have a rich historical presence, having been utilized in ancient artifacts long before the advent of modern nanotechnology. This section will weave together these themes, providing a comprehensive overview that sets the stage for deeper exploration into the remarkable potential of metallic nanostructures in shaping our future.

Chapter 1

Nanoscience

1.1 Miniaturization and Moore's Law

There has been a long-lasting trend to design and manufacture smaller mechanical, optical and electronic devices. This trend has been fueled by the computer industry, where an ongoing battle to design the smallest, fastest and cheapest integrated circuits has driven the miniaturization of transistors.

In 1965, Gordon Moore, while working for Fairchild Semiconductor, wrote an article for the 35th anniversary of *Electronics* magazine, where he noted that the circuit density of electronic components was doubling each year and anticipated that this rate would remain constant for at least 10 years (Moore, 1965); later, the integration pace was moderated to doubling approximately every 18 months (Shalf, 2015). This empirical observation, which led to Moore's prediction, lasted 40 years longer than he predicted and started to slow down when the fabrication technology of 2D lithography approached minimum feature sizes of just a few atoms.

In order to keep up with Moore's law, the minimum feature size of the fabrication process has to decrease with time. Photolithography, which is the current fabrication method for integrated circuits, is limited by diffraction, which depends on the wavelength of the light source.

Deep ultraviolet sources can be used to fabricate devices with sub-100 nm feature sizes; however, for feature sizes around 10 nm or less, novel nanofabrication techniques have to be implemented in the electronics industry.

1.2   Richard Feynman: *"There's Plenty of Room at the Bottom"*

Richard Phillips Feynman was a renowned physicist born in Queens, New York City, in 1918. Among his many awards was the 1965 Nobel Prize in Physics for "fundamental work in quantum electrodynamics, with deep consequences for the physics of elementary particles" (PhysicsNobel, 1965).

On December 29th of 1959, Richard Feynman gave a talk at the annual meeting of the American Physical Society at the California Institute of Technology (Caltech). This talk was titled "There's plenty of room at the bottom" (Feynman, 1960). This event is considered the birth of nanotechnology as a scientific field.

In this talk, Professor Feynman addressed the complaint students usually made that there was nothing left to discover in the field of physics. He mentioned the fields of low temperature and low/high pressure, where there were new developments in fundamental physics being discovered constantly, and he then introduced the problem of "manipulating and controlling things on a small scale," which would have a great number of technical applications.

As examples of things that could be done by manipulating things at a small scale, Feynman saw feasible to write a whole encyclopedia on the head of a pin. In order to make these things possible, manipulation at an atomic level should be possible. A key component to achieve this is the electron microscope. Feynman suggested that in order to make these things possible, the current electron microscope should improve its resolution at least 100 times. In 1960, the resolution of the electron microscope was around

10 Å; a resolution of 0.1 Å would allow us to see individual atoms. Current electron microscopes have resolutions which go below 0.5 Å, which is more than 20 times better than the resolution of microscopes in 1960. These electron microscopes which include current aberration correction techniques are capable of imaging individual atoms.

Figure 1.1 shows atomic resolution micrographs of $SrTiO_3$ taken with a ThermoScientific Titan, operated at an acceleration voltage of 300 kV, which, in the aberration-corrected scanning transmission electron microscope (STEM) mode, can achieve sub-Angstrom resolution (around 0.7 Å).

The possibility of manipulating things at the atomic scale would allow the fabrication of nanometer-sized devices, which would have different physical properties from their larger-size equivalents. These

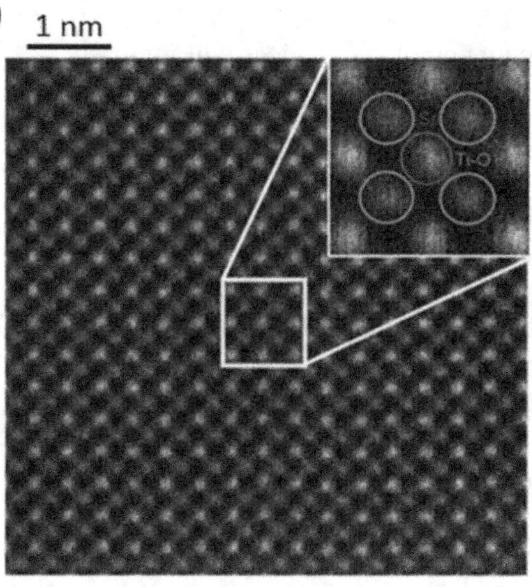

Fig. 1.1. Atomic resolution images of $SrTiO_3$. Reprinted from (Fatermans, 2018).

devices would be ruled by quantum mechanic laws, increasing both electrical resistance and friction in the case of mechanical systems. The possibility of fabricating nanodevices would allow investigating new forces and effects that occur at atomic scales.

Regarding the fabrication of devices on a small scale, Feynman suggested evaporation for the deposit of materials, including metals, and nanolithography for pattern generation; both techniques are now commonly used for the fabrication of nanostructures, and some of these fabrication techniques will be covered in Chapter 2 of this book.

Feynman in his talk suggested that nanostructures made of metallic materials could be used to explore inductors and capacitors at small scales, as well as antennas which could be able to emit and receive light. These devices could be used to fabricate circuits at small scales. Feynman also anticipated problems in building metallic structures at nanometer scales, one of which would be the electrical resistance of the structures. Also, by reducing the size of the devices, the natural frequency of the devices will increase.

1.3   Size Dependence of Physical Properties

The physical, chemical and biological properties of matter change as a function of size; this is due to the increase in surface-to-volume ratio when the size of a particle decreases. When a particle decreases in size, there is a higher percentage of atoms in its surface compared to the total number of atoms in the whole volume. The atoms in the surface are the ones that react with nearby atoms. The increase in the surface-to-volume ratio can be easily visualized by calculating the surface and volume of simple geometrical shapes. In the case of a cube, for example, the surface of the cube is given by $S = 6 \times l^2$, and its volume can be calculated using $V = l^3$, where l is the length of one of its edges. For a cube with the length of one of its edges equal to $1m$, its surface will be equal to $6m^2$ and its volume will be equal to $1m^3$, which would give a surface-to-volume ratio of $6\frac{1}{m}$. If we now

Table 1.1. Surface-to-volume ratio of cubes of different dimensions.

Length (m)	Surface/Volume ratio (1/m)
1 (1×10^0)	6
0.5	12
0.25	24
0.1 (1×10^{-1})	60
0.001 (1×10^{-3})	6000
0.000001 (1×10^{-6})	6×10^6
0.000000001 (1×10^{-9})	6×10^9

divide this cube into smaller cubes with edge length $l = 0.5m$, we will have 8 cubes with a total surface of $8 \times 6 \times (0.5m)^2 = 12m^2$, and the volume for the 8 cubes is still $1m^3$, but now the surface-to-volume ratio is $12\frac{1}{m}$.

Table 1.1 shows the surface-to-volume ratio for a cube as a function of edge length. For edge lengths in the nanometer range, the surface is 10^9 times larger than the volume, which indicates that a nanometer-sized particle will be 10^9 more reactive per unit volume than a meter-sized particle. Also, reducing the particle size to the nanometer range will increase its interaction with light in the visible spectrum. The size of the particle will be comparable to the wavelength of visible light, and the optical properties of the material will change. This can be seen with gold, for example: a large gold particle will reflect yellow light, while gold nanoparticles in a solution will scatter red light.

The properties of nanoparticles also change as a function of shape, which is due to the difference in the surface-to-volume ratio of different nanoparticles. Figure 1.2 shows the surface-to-volume ratio of cubes, spheres and square pyramids as a function of their edge length, diameter and base edge length, respectively. In the graph, it can be seen how the surface-to-volume ratio is different depending on the shape of the particle. This varies the physical properties of the particle for a fixed volume particle.

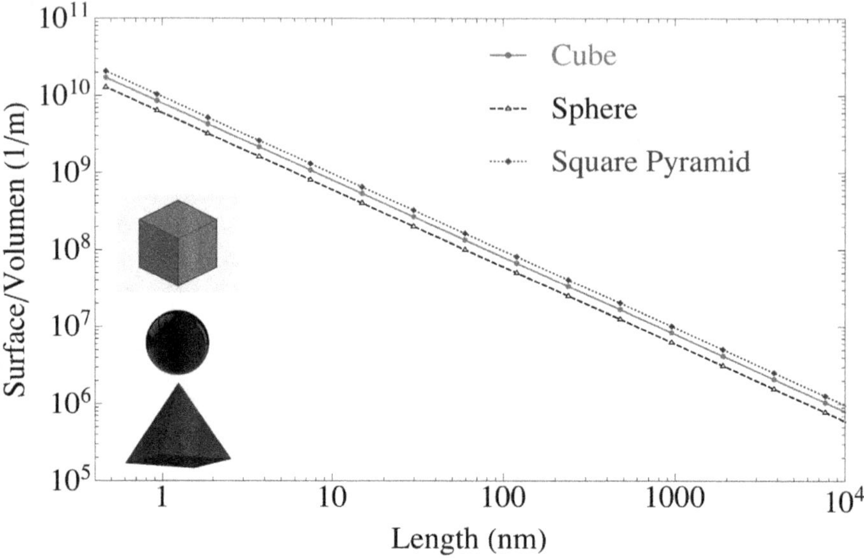

Fig. 1.2. Surface-to-volume ratio of different particle shapes as a function their main length.

1.4 Metallic Nanoparticles in Ancient Materials

Metallic nanoparticles have been used since ancient times in colored window glass, ceramics and even as colorants for clothing. Metallic nanoparticles have been popular as colorants due to their strong scattering properties in the visible range that result in very intense and attractive colors. These colored materials that included metallic nanoparticles were discovered empirically by craftsmen, who noticed that bright colors emerged from ordinary materials when they were combined with metallic oxide powders. In the case of stained glass, if these metallic oxide powders were added during the melting process, the resulting glass would have specific colors depending on the metal powder used. Copper oxides gave a green or bluish-green color, while cobalt resulted in deep blue and gold in bright red (Nayfeh, 2018).

These empirical findings were transferred among craftsmen through generations without knowledge of the effect of metallic

Fig. 1.3. Lycurgus cup.

nanoparticles on the final result. Famous examples of ancient applications of metallic nanoparticles include the Lycurgus cup, a Roman glass cup dating back to the 4th century and currently exhibited in the British Museum. This cup is an example of dicroic glass, where nanoparticles of gold and silver of around 50–70 nm in diameter are distributed inside the glass, which generates a green reflection of light and a red transmission. Figure 1.3 shows the reflection and transmission properties of the Lycurgus cup.

Part 2

Physical Properties

In this part, we delve into the intriguing physical properties of metallic nanostructures, a cornerstone of their remarkable versatility and functionality. As we explore the optical, electrical, and thermal characteristics of these materials, we will uncover how their behavior at the nanoscale differs significantly from that of their bulk counterparts. This unique size-dependent behavior not only enhances their performance in various applications but also opens new avenues for innovation in fields ranging from electronics to photonics. We will begin by examining the optical properties that allow metallic nanostructures to manipulate light in unprecedented ways, leading to advancements in imaging and sensing technologies. Next, we will explore their electrical properties, which are crucial for the development of next-generation electronic devices, including transistors and sensors. Finally, we will discuss the thermal properties that govern heat transfer and management in nanoscale systems, highlighting their importance in energy applications.

Chapter 2

Optical Properties of Metallic Nanostructures

The description of light matter interactions at the nanoscale can be done most accurately using quantum electrodynamics, however a good approximation can be achieved using classical electromagnetic theory of fields (Zuloaga, 2009). The use of classical electromagnetic theory is justified in explaining experimental results since in general laboratory equipment will not be sensitive to individual photons, which is the detail that quantum electrodynamics can achieve. In an experiment, the cumulative effect of a large amount of photons will appear as a continuous macroscopically observable response, which can be described by classical electromagnetic theory.

2.1 Classical Electromagnetic Theory of Fields

From 1861 to 1865, James Clerk Maxwell, Oliver Heaviside, and Josiah Willard Gibbs developed modern electromagnetic wave theory which was experimentally verified by Heinrich Hertz in 1887 (Li, 2017). Maxwell gathered four empirical equations that were formulated by Gauss, Ampère and Faraday and modified Ampère's Law to include a term for the displacement current, with this addition the electric and magnetic field equations could be solved into a wave equation with a velocity term equal to the speed of light, this finding led to the discovery that light is an electromagnetic wave.

Maxwell's equations:

$$\nabla \cdot \vec{E} = \frac{\rho}{\epsilon}, \tag{2.1a}$$

$$\nabla \cdot \vec{B} = 0, \tag{2.1b}$$

$$\nabla \times \vec{E} = -\frac{\partial \vec{B}}{\partial t}, \tag{2.1c}$$

$$\nabla \times \vec{B} = \mu \vec{J} + \mu \epsilon \frac{\partial \vec{E}}{\partial t}, \tag{2.1d}$$

where $\vec{E}$ is the electric field in $[V/m]$, $\vec{B}$ is the magnetic flux density in $[Wb/m^2]$ or $[T]$, ϵ and μ are the permittivity and permeability of the medium, ρ is the total electrical charge density in $[C/m^3]$ and $\vec{J}$ is the external current density in $[A/m^2]$.

Since light is an electromagnetic wave composed of an electric field and a magnetic field, then the interaction of light with matter is dependent on the permeability and permittivity of the material, this interaction depends on the frequency of the electromagnetic wave since materials in general are dispersive, which means their properties change with frequency.

The constitutive equations in electromagnetic wave theory describe the electrical and magnetical response of a medium, they relate the displacement field $\vec{D}$ with the electric field $\vec{E}$, the magnetic induction $\vec{B}$ with the magnetic field $\vec{H}$ and the electrical current $\vec{J}$ to the electric field through the following equations:

$$\vec{D}(\vec{r}, \omega) = \epsilon(\vec{r}, \omega)\vec{E}(\vec{r}, \omega), \tag{2.2a}$$

$$\vec{B}(\vec{r}, \omega) = \mu(\vec{r}, \omega)\vec{H}(\vec{r}, \omega), \tag{2.2b}$$

$$\vec{J} = \sigma(\vec{r}, \omega)\vec{E}(\vec{r}, \omega). \tag{2.2c}$$

The permittivity $\epsilon(\vec{r}, \omega)$ and permeability $\mu(\vec{r}, \omega)$ of the medium account for the displaced charge and magnetization of the medium due to an electric and magnetic fields, respectively. Both permittivity and permeability can change with the position, intensity and frequency of the field, and this is true for inhomogeneous, anisotropic, nonlinear and dispersive materials.

In the case of homogeneous materials, both the permittivity and permeability can be represented as constant scalars, while tensors are needed to describe anisotropic materials.

Both the displaced charge and magnetization of matter due to electric and magnetic fields do not happen instantly, which can be represented by a phase difference between these vectors. The change in phase is included by making the permittivity, permeability and conductivity a complex function.

At optical frequencies, the complex electric permittivity $\epsilon = \epsilon' + i\epsilon''$ of a nonmagnetic medium can be determined experimentally measuring the optical properties of metals, such as the complex refractive index:

$$\bar{n} = \sqrt{\epsilon} = n + i\kappa, \tag{2.3a}$$

where n is the refractive index and κ is the extinction coefficient which determines the optical absorption (Klimov, 2014).

From (2.3a), the following relations can be obtained:

$$\epsilon' = n^2 - \kappa^2, \tag{2.4a}$$

$$\epsilon'' = 2n\kappa, \tag{2.4b}$$

$$n^2 = \frac{\epsilon'}{2} + \frac{1}{2}\sqrt{\epsilon'^2 + \epsilon''^2}, \tag{2.4c}$$

$$\kappa^2 = \frac{\epsilon''}{2n}. \tag{2.4d}$$

The extinction coefficient κ is related to the absorption coefficient in the Beer–Lambert law $I(x) = I_o e^{-\alpha x}$ by the relation

$$\alpha(\omega) = \frac{2\kappa(\omega)\omega}{c}, \tag{2.5a}$$

which means that the imaginary part of the electrical permittivity defines the absorption as a function of frequency of a medium.

2.2   The Drude–Sommerfeld–Lorentz Theory of Optical Properties of Metals

The optical properties of metals can be modeled by considering conduction taking place as a gas of free electrons moving in a positively charged crystal lattice. For alkaline metals, this model extends to the ultraviolet region, while for noble metals, interband transitions which take place in the visible region restrict the accuracy of this model.

In this free electron model, also known as the plasma model, electron–electron interactions and defects in the crystal lattice are ignored, the effective electron mass is also used instead of the free electron mass. Using this model, the motion of electrons due to an external electric field is given by the following equation:

$$m * \left(\frac{d^2 x}{dt^2} + \gamma \frac{dx}{dt} \right) = -e\vec{E}(t), \tag{2.6a}$$

where $\gamma = 1/\tau$ is the damping frequency which corresponds to the frequency of electron collisions. The value τ is the relaxation time of the free electron gas and is in the $10^{-14}s$ order of magnitude.

If we consider an external harmonic monochromatic electric field $E(t) = E_0 e^{-i\omega t}$, like the one present in an optical plane wave, for example, the solution to Eq. (2.6a) has the form

$$x(t) = \frac{e}{m * (\omega^2 + i\omega\gamma)} E(t). \tag{2.7a}$$

The displacement $x(t)$ of the electron generates a dipole moment and a polarization of the medium $\vec{P}$ given by

$$\vec{P} = -n_0 e\vec{x} = -\frac{n_0 e^2}{m * (\omega^2 + i\omega\gamma)} \vec{E}, \tag{2.8a}$$

where n_0 is the concentration of conduction electrons.

From (2.8a), the susceptibility κ of the metal is defined by the relation

$$\kappa = -\frac{n_0 e^2}{m*(\omega^2 + i\omega\gamma)},$$ (2.9a)

and the electric permittivity ϵ is given by

$$\epsilon(\omega) = 1 - \frac{\omega_{pl}^2}{\omega^2 + i\omega\gamma},$$ (2.10a)

where $\omega_{pl} = \sqrt{\frac{n_0 e^2}{\epsilon_0 m*}}$ is the plasma frequency of the free electron gas. The real and imaginary parts of the electric permittivity are given by

$$\epsilon' = 1 - \frac{\omega_{pl}^2 \tau^2}{1 + \omega^2 \tau^2},$$ (2.11a)

$$\epsilon'' = \frac{\omega_{pl}^2 \tau}{\omega(1 + \omega^2 \tau^2)}.$$ (2.11b)

It is worth noting that the complex part of the permittivity plays an important role at high frequencies and for noble metals.

The electrical conductivity is related to the permittivity through the following equation (Ingold, 2009):

$$\epsilon(\omega) = 1 + i\frac{\sigma(\omega)}{\omega}.$$ (2.12a)

Table 2.1 shows the Drude parameters for gold, silver and copper.

Table 2.1. Drude parameters for gold, silver and copper.

	Au	Ag	Cu
ω_{pl}, [eV]	9.1	9.1	8.8
τ, [fs]	29	40	40

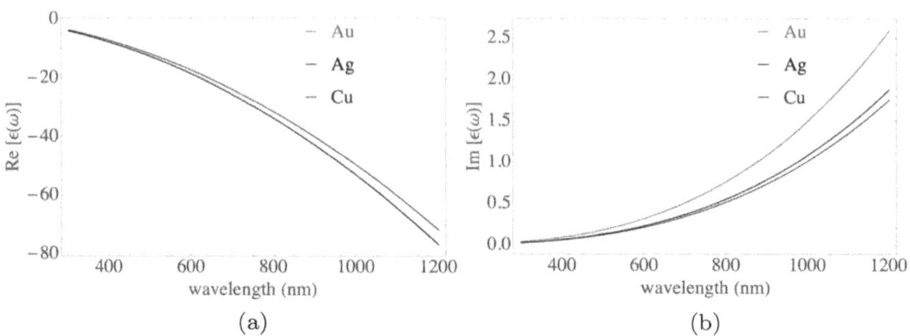

Fig. 2.1. (a) Real and (b) complex permittivity as a function of wavelength calculated from the Drude–Sommerfeld model.

By using the values in Table 2.1 and converting the photon energy to its equivalent wavelength using the equation $E = (hc)/\lambda$, where h is the Planck constant and c is the speed of light in vacuum, we can calculate the real and complex permittivity as a function of wavelength (Fig. 2.1).

From Fig. 2.1, we can see how the real permittivity is negative for metals and that at infrared wavelengths the imaginary part of the permittivity becomes relevant. The real permittivity of metals is due to the large number of free electrons which react with a lag with respect to the driving field.

There are two main mechanisms that contribute to the permittivity $\epsilon(\omega)$ in metals, the fast response of conduction electrons which can be modeled using the Drude–Sommerfeld theory and the electronic interband transitions which can be modeled using a Lorentz-like term (Giannini, 2011).

The Drude–Sommerfeld contribution can be written as Eq. (2.10a) with a term that accounts for the DC permittivity ϵ_∞, which is due to the residual polarization of the material:

$$\epsilon_{\text{Drude}}(\omega) = \epsilon_\infty - \frac{\omega_{pl}{}^2}{\omega^2 + i\omega\gamma}. \tag{2.13a}$$

The electronic interband transitions appear when photons with energy larger than the material bandgap are able to promote electrons from the valence band to the conduction band, this effect can be modeled using a Lorentz-like term

$$\epsilon_{\text{Lorentz}}(\omega) = \frac{\Delta\epsilon\Omega_p{}^2}{\Omega_p{}^2 - \omega^2 - i\Gamma\omega}, \qquad (2.14\text{a})$$

where, in analogy with the Drude formula, Ω_p and Γ are the plasma and damping frequencies for the bound electrons, and the parameter $\Delta\epsilon$ weighs the contribution of the interband transition to the permittivity (Giannini, 2011). Several of these Lorentz terms can be added to account for different interband transitions and get a better fit to experimental data.

Figure 2.2 shows the real and imaginary parts of the permittivity of gold calculated using the Drude–Sommerfeld and the Drude–Lorentz models and comparing the results to experimental values. From Fig. 2.2, it can be seen how for red and infrared wavelengths the permittivity is dominated by its large negative part, at this wavelength range the refractive index $n = \sqrt{\epsilon}$ is almost purely imaginary in this wavelength range. At shorter wavelengths, the presence of interband transitions is not explained by the Drude model, so a Lorentzian correction needs to be included. A single Lorentzian correction increases the range of the agreement with experimental data to $\lambda = 400$ nm, for shorter wavelengths agreement can be

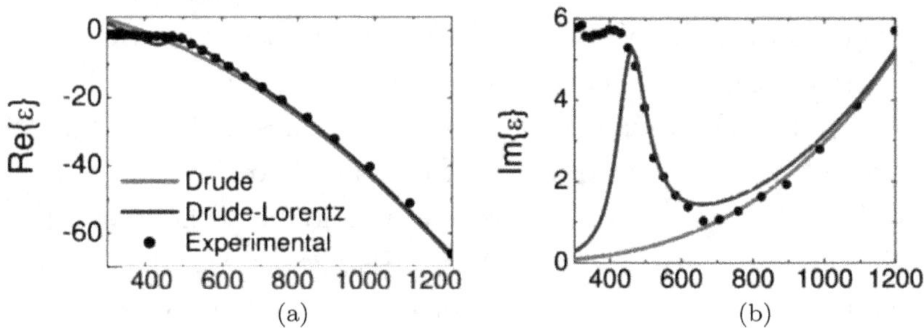

Fig. 2.2. (a) Real and (b) imaginary parts of the permittivity of gold as a function of wavelength calculated from the Drude–Sommerfeld model (red) and the Drude–Lorentz model (blue) as well as experimental data (black dots). Taken from (Giannini, 2011).

achieved by introducing more than one Lorentzian term (Giannini, 2011).

2.3 Optical Properties of Metals

Optical properties of materials are fundamental to develop accurate optical models, and for computer simulations, these optical properties can be found in the literature for microwave and millimeter frequencies and for optical frequencies in the visible region, however the optical properties of metals vary significantly in the infrared and terahertz regions of the electromagnetic spectrum. For metallic nanostructures working in the infrared and terahertz regions, the optical properties have to be determined experimentally using methods like ellipsometry (Gonzalez, 2009). Figure 2.2 shows the optical parameters n and k for different metals at terahertz and infrared frequencies (see Gonzalez, 2009).

2.4 Light Interaction with Metallic Nanoparticles

2.4.1 *Analytical Solutions*

The first theoretical work on the scattering of light by particles smaller than the incident wavelength was performed by Lord Rayleigh at the end of the 19th century (Nayfeh, 2018). Lord Rayleigh (born John William Strutt) was a British mathematician that first described the elastic scattering of light by small particles, these objects include oxygen and water molecules of the sky. Rayleigh discovered that the scattering is proportional to the sixth power of the object's diameter and inversely proportional to the fourth power of the wavelength (i.e. $I \propto d^6, \lambda^{-4}$). His theory explained physical phenomena, such as the blueness of the sky and the redness of the sunset (Nayfeh, 2018).

Gustav Mie derived the solution of Maxwell's equations for spherical particles with arbitrary radius and composition and for infinitely long cylinders. Based on the results of Rayleigh and Mie, Gans expanded the solution to elliptical particles, which became known as the Mie–Gans theory. Gans demonstrated that the optical reponse of metallic nanoparticles is strongly shape-dependent and explained the origin of the different colors produced when white light is incident on colloidal solutions of elliptically shaped nanoparticles with different aspect ratios (Giannini, 2011).

A formal solution of rigorous scattering theory for nanoparticles is unfortunately only possible for restricted geometries. In most practical applications, it is desired to work with arbitrary shaped nanostructures and take advantage of certain structure- dependent qualities, such as hot spots, in these cases a computational approach to solve Maxwell's equations is used.

The interaction of light with metallic nanostructures has led to a new branch of photonics known as plasmonics. Surface plasmons are groups of electrons collectively moving in a harmonic way at the interface between two materials which possess respectively a positive and a negative real part of their permittivity, such as a metal/dielectric interface. Surface plasmons confine light to very small dimensions which enhances the light–matter interaction, resulting in very unique properties and applications. Surface plasmons can be divided in two categories: localized and propagating surface plasmon polaritons (SPPs).

Localized surface plasmons (LSPs) appear in metallic nanoparticles of size comparable to or smaller than the excitation wavelength, incident light on a metallic nanoparticle causes a coherent oscillation of conduction electrons that greatly enhances the electric field near the particle's surface, this coherent oscillation also induces an optical absorption which has a maximum at the plasmon resonant frequency. The surface plasmon resonance can be tuned based on the shape of the nanoparticle and the metal permittivity, the enhanced electric field induced falls off significantly with distance from the surface of the nanoparticle, in the case of noble metal nanoparticles the plasmon resonance occurs at visible wavelengths which generates brilliant colors in metal colloidal solutions.

Surface plasmon polaritons are electromagnetic waves that travel in a metal/dielectric interface losing energy to the metal due to absorption. These SPPs occur when both media have frequency-dependent complex permittivities with real part of opposite signs.

The first observation of plasmons date back to 1902 when Robert Williams Wood noticed narrow dark bands in the diffraction spectrum of an illuminated metallic grating, these dark bands are known as Wood's anomalies. Ugo Fano associated these anomalies with the excitation of electromagnetic surface waves on the diffraction gratings.

A major boost to the field of plasmonics came when Martin Flashmann and coworkers observed strong Raman scattering from pyridine molecules in the vicinity of toughened silver surfaces (Kik, 2007), which led to the now well-established field of surface enhanced Raman spectroscopy (SERS).

2.5 Localized Surface Plasmons in Different Materials

Surface plasmons appear when photons interact with free electrons, therefore they have been investigated using materials that have a large number of free electrons, like metals, noble metals such as gold and silver have a large amount of free electrons and their resonance wavelength is located in the visible and near-infrared regions, making them good experimental materials for LSPs. However, LSPs can also appear in heavily doped semiconductors and 2D materials. The plasmon frequency ω_p is given by

$$\omega_p = \sqrt{\frac{Ne^2}{\epsilon_0 m^*}}, \tag{2.15}$$

where N is the density of carriers (electrons or holes), m^* is the effective mass of carriers and ϵ_0 is the permittivity of vacuum. Figure 2.3

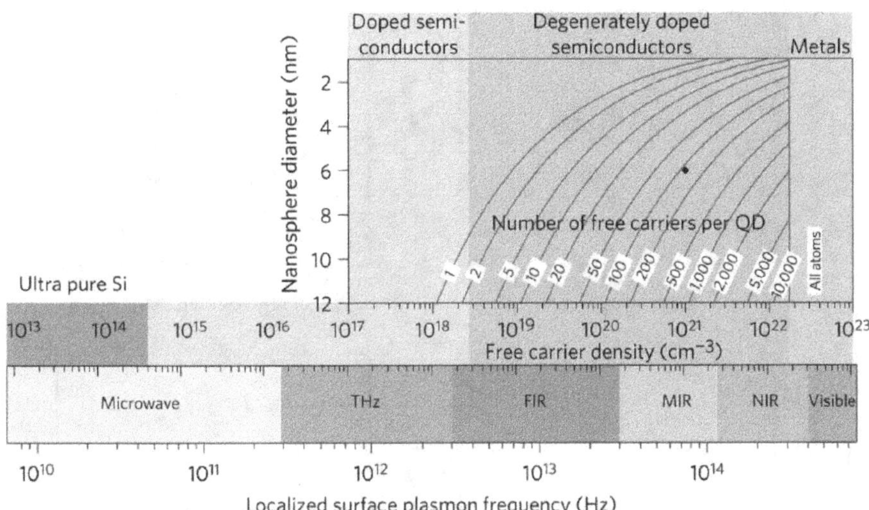

Fig. 2.3. Plasmon frequency dependence as a function of carrier density for metals and semiconductors. Taken from (Yu, 2019).

shows the value of ω_p as a function of carrier concentration, for noble metals a typical value of N is $10^{22} - 10^{23}$ cm^{-3}, making the plasmon resonance wavelength appear in the NIR and visible region, for doped semiconductors the typical value of N is $10^{16} - 10^{21}$ cm^{-3}, those materials have a plasmon resonance in the THZ and FIR regions of the electromagnetic spectrum. From Fig. 2.3, it can be seen that the plasmon frequency can be tuned by changing the doping concentration.

Besides the material properties, the plasmon resonance also depends on the size, shape and placement of the structures or array of structures, Fig. 2.4 shows changes in the plasmon resonance due to the shape of nanostructures of the same material, silver.

Plasmon resonances have also been observed in 2D materials, graphene, for example, which has ultrahigh carrier mobility present a plasmon resonance in the infrared and THz regions.

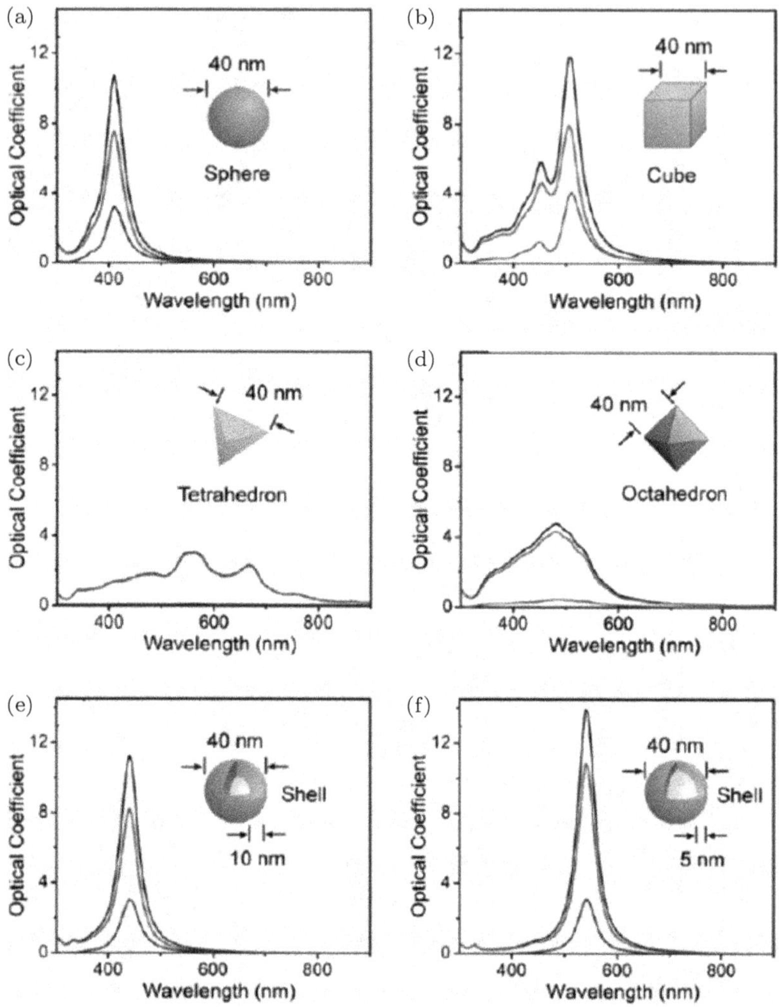

Fig. 2.4. Calculated extinction coefficient (black), absorption (red) and scattering (blue) for solid and hollow silver nanoparticles. Taken from (Wiley, 2006).

2.6 Derivation of Surface Plasmon Polaritons

Plasmons are (bosonic) elementary excitations in a metallic solid (Trügler, 2016). To generate these excitations, a movable elementary charge has to be displaced by an external field and a restoring force has to be restored to start an oscillation. The displacement of an electron due to an electromagnetic field can be modeled by the Drude–Sommerfeld theory and is given by Eq. (2.7a). From the Drude–Sommerfeld model, the electric permittivity is given by (2.10a), neglecting γ the Drude permittivity simplifies to $1 - \omega_{pl}^2/\omega^2$, which is positive for $\omega > \omega_{pl}$ and negative for $\omega < \omega_{pl}$, a negative permittivity will give an imaginary refractive index, while a positive permitivity results in a real refractive index. An imaginary refractive index implies that an electromagnetic wave cannot propagate inside the medium, it will be absorbed to a given depth known as the skin depth.

These surface plasmon polaritons, or SPPs, are light waves trapped on the surface because of their interaction with free electrons of the conductor, the resonant interaction between the surface charge oscillation and the electromagnetic field of the light gives rise to the unique properties of SPPs.

A remarkable property of SPPs is their ability to keep optical energy concentrated at the nanoscale and channel it by using subwavelength structures. This could lead to miniaturized photonic circuits (Barnes, 2003). These circuits would first convert light into SPPs and then propagate and process it before converting them back to light.

The interaction between the surface charges and the electromagnetic wave results in the momentum of the SPP mode $\hbar k_S P$ being greater than that of a free-space photon of the same frequency ($\hbar k_0$.) Solving Maxwell's equations yields the SP dispersion relation (Barnes, 2003)

$$k_S P = k_0 \sqrt{\frac{\epsilon_d \epsilon_m}{\epsilon_d + \epsilon_m}}. \tag{2.16}$$

The frequency-dependent permittivity of the metal, ϵ_m, and the dielectric material, ϵ_d, must have opposite signs if the SPP is possible at such an interface. This condition is satisfied for metals because ϵ_m is both negative and complex.

Another characteristic of SPP is that they only propagate along the surface, the field perpendicular to the surface decays exponentially with distance from the surface through an evanescent wave.

In general, a good plasmonic material can be identified if the following properties are met:

$$Re\{\epsilon_m\} < 0 \quad \text{and} \quad Im\{\epsilon_m\} \ll -Re\{\epsilon_m\}. \tag{2.17}$$

2.7 Excitation of Surface Plasmons

There are several techniques to generate SPPs, they can be generated by using a prism, by scattering from a topological defect on the surface, such as a subwavelength protrusion or hole or by a periodic corrugation in the metal's surface. It is worth noting that the reverse process also allows an SPP mode to couple with light with good efficiency.

The excitation of a surface plasmon becomes possible when the photon momentum component parallel to the surface is increased by some means, this will generate a pure real propagation vector along the interface and a purely imaginary propagation vector perpendicular to the interface, this will give a propagating wave along the interface and an evanescent wave perpendicular to the interface.

As discussed before, a plasmon can be generated by attenuated total internal reflection, with a surface diffraction grating or with a nanolocalized light source. Figure 2.5(a) shows plasmon generation using the Otto configuration, in this configuration the total internal reflection at the prism/air interface generates an evanescent field that excites an SPP at the dielectric/metal interface, the distance between the metal and the prism should be in the order of λ. In Fig. 2.5(b), the total reflection at the prism/metal interface generates an evanescent field that excites a surface plasmon at the opposite metal/air interface, the thickness of the metal film must be smaller than the skin depth, this configuration is known as the Kretschmann configuration.

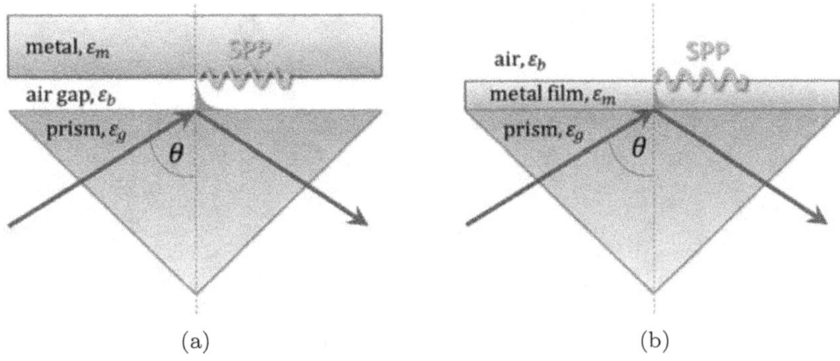

(a) (b)

Fig. 2.5. Two configurations that provide the missing momentum contribution discussed in the text for the excitation of surface plasmons: (a) Otto configuration; (b) Kretschmann configuration. Taken from (Trügler, 2016).

2.8   Surface Plasmon Propagation

Once light has been converted to an SPP, it will propagate along a flat metal surface but will be attenuated due to the absorption in the metal. The attenuation will depend on the permittivity of the metal at the frequency of the SPP. The propagation length can be calculated using the imaginary part of the complex surface plasmon wavevector, $k_{\mathrm{SPP}} = k'_{\mathrm{SPP}} + ik''_{\mathrm{SPP}}$ from the SPP dispersion equation

$$\delta_{\mathrm{SPP}} = \frac{1}{2k''_{\mathrm{SPP}}} = \frac{c}{\omega}\left(\frac{\epsilon'_m + \epsilon_d}{\epsilon_d \epsilon'_m}\right)^{3/2}\frac{(\epsilon'_m)^2}{\epsilon''_m}, \tag{2.18}$$

where ϵ'_m and ϵ''_m are the real and imaginary parts of the permittivity of the metal.

In the past, the SPP propagation length was smaller than the size for the components at the time, with current nanofabrication capabilities, it is now possible to fabricate components that are smaller than the propagation length, these components could be integrated into circuits before propagation losses become too significant (Barnes, 2003).

Chapter 3

Electrical Properties

When decreasing the size of metallic structures, the surface area per unit volume increases, which induces a change in their mechanical, thermodynamic, electrical and optical properties (Lacy, 2011).

3.1 Resistivity

When reducing one of the dimensions of a conducting material the electrical resistivity changes from its bulk values. This change occurs because the mean free path of conduction electrons is reduced due to an increase in scattering effects (Lacy, 2011).

In order to model the electrical resistivity of metallic nanostructures, a model for the scattering mechanisms has to be used. The scattering mechanisms are due to the surface of the material as well as scattering from grain boundaries, uneven or rough surfaces and scattering due to impurities (Lacy, 2011). Most of these scattering mechanisms are dependent on the fabrication methods and conditions, so they are difficult to quantify; however, the increase in scattering will reduce the mean free path of electrons, increasing the electrical resistivity.

A model that relates the electrical resistivity to film thickness and temperature was developed by Fred Lacy (Lacy, 2011) and can be expressed as two equations: one that applies to the case when the thickness of the film is larger than twice the mean free path of the conduction electrons in that material (l_{bulk}) (Eq. (3.1)) and one for the case when the thickness is smaller than twice l_{bulk} (Eq. (3.2)). This resistivity model is a good fit to experimental results (Lacy, 2011):

$$\rho = c\rho_0 \left[\frac{1}{\frac{\left[2\sqrt[\delta]{\gamma/kT}-b\right](\tau_1/\tau_2-1)}{2a+b} + 1} \right] \quad \text{if} \quad t \geq 2l_{\text{bulk}} \qquad (3.1)$$

or

$$\rho = \frac{c\rho_0}{\kappa'[1-\ln\kappa']} \left[\frac{1}{\frac{\left[2\sqrt[\delta]{\gamma/kT}-b\right](\tau_1/\tau_2-1)}{2a+b} + 1} \right] \quad \text{if} \quad t \leq 2l_{\text{bulk}}, \qquad (3.2)$$

where $\rho_0 = \frac{m}{ne^2\tau_2}$, a is the atomic radius, b is the size of the opening between atoms (when the atoms are stationary), τ_1/τ_2 is the ratio of travel time before scattering, δ is 1, k is the Boltzmann constant, T is the temperature in Kelvin, and γ is the proportionality constant in the energy equation $U = \gamma/r^\delta$ (Lacy, 2011).

At room temperature, equations (3.1) and (3.2) reduce to the following equations:

$$\rho = \rho_0' = c\rho_0, \qquad (3.3)$$

where ρ_0 is the bulk resistivity for the material, c is a constant ($c \geq 1$), and ρ_0' is the modified bulk resistivity due to the additional scattering effects;

$$\rho = \rho_0' = \frac{c\rho_0}{\kappa'[1-\ln\kappa']}, \qquad (3.4)$$

where κ' is a constant that depends on an effective thickness $t' = t-\eta$, where t is the film thickness and η is the reduction factor, as a result of which this constant can be written as $\kappa' = (t-\eta)/(2l_{\text{bulk}})$.

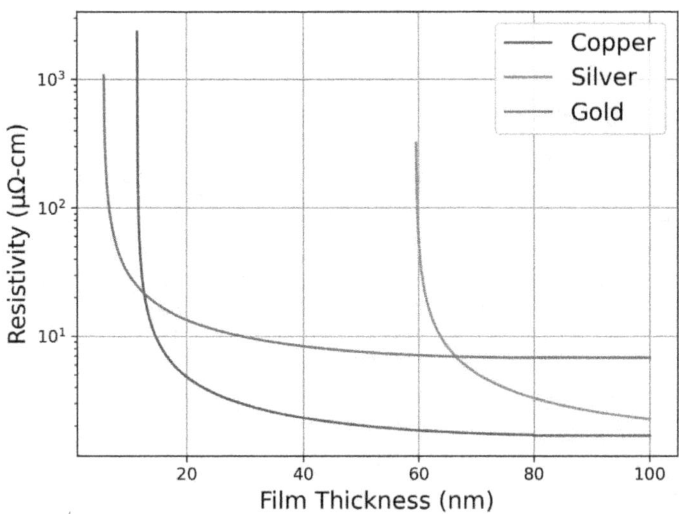

Fig. 3.1. Calculated resistivity for Cu, Ag and Au as a function of film thickness.

Table 3.1. Values used to calculate resistivity. Taken from (Gall, 2016).

Metal	Bulk Resistivity	Mean Free Path	η	c
Copper	1.678×10^{-8}	3.99×10^{-8}	11.4	1
Silver	1.587×10^{-8}	5.33×10^{-8}	59.5	1.06
Gold	2.214×10^{-8}	3.77×10^{-8}	5.5	3.07

Figure 3.1 shows the calculated resistivity for copper, silver, and gold as a function of film thickness. The calculations were made using Eqs. (3.3) and (3.4) and using the values shown in Table 3.1. These values were obtained from (Gall, 2016).

Chapter 4

Thermal Properties

In previous chapters, we have seen that the surface-to-volume ratio changes the physical properties of particles. In the case of nanoparticles, the surface-to-volume ratio can be in the $10^9 - 10^{10}$ range, which means that the number of surface atoms of a material can be $10^9 - 10^{10}$ higher for the same volume made from nanoparticles compared to the same volume of bulk material.

4.1 Melting Temperature of Metallic Nanoparticles

One of the many differences in the physical properties of a nanostructured material is that its melting temperature is lower than the same material in bulk. The shape of the nanomaterial also influences the melting temperature since the ratio of the number of atoms in the surface changes with the shape of the material (1.2).

By using classical thermodynamics and applying the conservation law for a nanostructure at the melting phase transition, the melting temperature of a nanostructure can be obtained using (Guisbiers, 2008)

$$T_m = T_{m,\infty} \left[1 + \frac{(\gamma_1 - \gamma_s)}{\Delta H_{m,\infty}} \frac{A}{V} \right], \qquad (4.1)$$

where $T_{m,\infty}$ is the bulk melting temperature of the material in K, γ_1 and γ_s are the surface tensions in the liquid and solid phases, respectively, in J/m^2, $\Delta H_{m,\infty}$ is the bulk melting enthalpy in J/m^3, and $\frac{A}{V}$ is the surface-to-volume ratio of the nanoparticles.

In order to simplify Eq. (4.1), Guisbers *et al.* (Guisbiers, 2008) introduced a shape parameter α_{shape}:

$$\alpha_{\text{shape}} = 2 * A * L * \frac{\gamma_s - \gamma_1}{V * \Delta H_{m,\infty}}, \tag{4.2}$$

where the parameter L represents the length size of the nanostructure, i.e., the radius for the case of a spherical structure or the length side for non-spherical structures. Using this shape parameter, Eq. (4.1) can be rewritten as

$$T_m = T_{m,\infty} \left(1 - \frac{\alpha_{\text{shape}}}{2L}\right), \tag{4.3}$$

Table 4.1 shows some parameters taken from (Guisbiers, 2008) to calculate the melting point of different nanoparticles as a function of shape and size.

Figures 4.1, 4.2 and 4.3 show the calculated melting points for gold, silver and copper as a function of nanoparticle size and shape. The calculations were made using Eq. (4.3) and the values shown in Table 4.1.

Table 4.1. Values used to calculate resistivity. Taken from (Guisbiers, 2008).

Metal	$T_{m,\infty}$ (K)	α_{sphere} (nm)	$\alpha_{\text{tetrahedron}}$ (nm)	α_{cubic} (nm)
Copper	1358	1.63	7.98	3.26
Silver	1235	1.88	9.22	3.76
Gold	1337	1.83	8.96	3.66

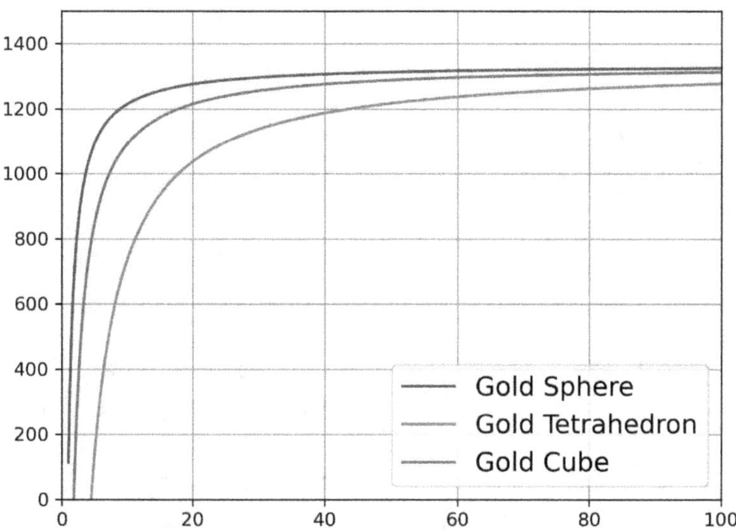

Fig. 4.1. Calculated melting point of gold nanoparticles as a function of nanoparticle size and shape.

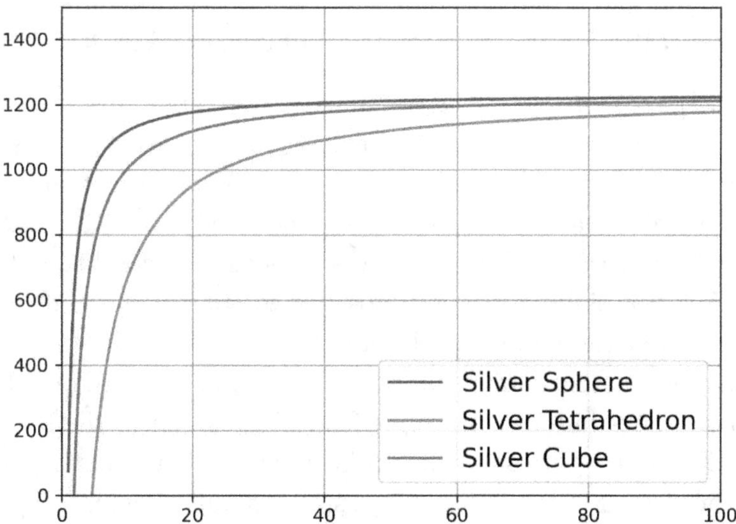

Fig. 4.2. Calculated melting point of silver nanoparticles as a function of nanoparticle size and shape.

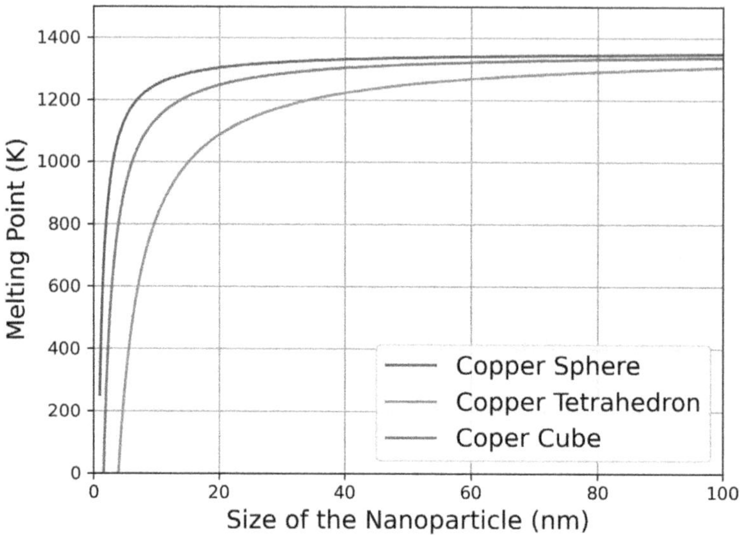

Fig. 4.3. Calculated melting point of copper nanoparticles as a function of nanoparticle size and shape.

4.2 Metallic Nanostructures as Sources of Heat

Metallic nanostructures can be tuned to enhance absorption and scattering at specific wavelengths through localized surface plasmons. A side effect of increased absorption is an increase in temperature of the metallic nanostructure. This "side effect" can turn metallic nanoparticles into ideal nano-sources of heat that can be controlled remotely using light, making it possible to control thermal-induced phenomena at the nanoscale (Baffou, 2013).

Among different nanoparticles, the ones made out of metal show a stronger light-induced heating effect. This is due to the high concentration of free carriers, which collectively oscillate under plasmon resonance, generating mostly heat since metallic nanoparticles are very poor light emitters.

The heating effect can be strongly enhanced by coupling different metallic nanoparticles. There are two main interaction mechanisms that can enhance the heating effect: (i) addition effect and (ii) Coulomb interaction.

The addition effect comes from combining the heat flux generated by single nanoparticles through the thermal diffusion equation. The Coulomb interaction between nanoparticles originates from the plasmonic increase in electric field due to incident electromagnetic radiation. This increase in electric field can interfere constructively when the nanoparticles are placed at a certain distance from each other. This electromagnetic constructive interference will increase the heat generation at each nanoparticle (Govorov, 2007).

Part 3

Synthesis

In this part, we turn our attention to the synthesis of metallic nanostructures, a critical aspect that bridges the gap between theoretical understanding and practical application. The ability to create these structures with precision and control is essential for harnessing their unique properties and unlocking their potential across various fields. We will explore three prominent methods of synthesis: self-assembly, lithography and silicon micromachining. Self-assembly techniques, inspired by natural processes, allow for the spontaneous organization of nanoparticles into complex structures, offering a cost-effective and scalable approach. Lithography, on the other hand, provides the precision needed to fabricate intricate patterns at the nanoscale, enabling advancements in electronics and photonics. Finally, silicon micromachining combines traditional semiconductor fabrication techniques with nanotechnology, paving the way for innovative applications in sensors and microelectromechanical systems.

Chapter 5

Fabrication

5.1 Self-Assembled Metallic Nanostructures

Metallic nanostructures can be fabricated using a "top-down" or a "bottom-up" approach. In the 'bottom-up approach, individual atoms and molecules are brought together or self-assembled to form the desired structure.

The top-down approach doesn't involve chemical synthesis or self-assembly, and it is usually an extension of the techniques used for microfabrication. In the top-down approach, the material is etched or selectively deposited, usually employing nanolithography to define a nanometer-sized pattern, and then evaporating or sputtering material on top of that pattern.

The bottom-up fabrication process can be traced back to the 1850s when Michael Faraday synthesized his famous gold colloids; however, only recently, the controllability over size, shape, structure and composition of metallic nanostructures reached the level required for realistic applications (Giannini, 2011).

There are multiple techniques for the bottom-up fabrication of metallic nanostructures. The reduction of metal salts by a chemical agent or by a photochemical process can be used to produce a large number of nanoparticles with very good control over their size and shape. This technique has been used to produce metallic spheres, cubes, rods, octahedrons, triangular prisms, boxes and stars (Giannini, 2011).

Bottom-up techniques give very fine nanostructures of individual nanoparticles, nanoshells or nanowires, with narrow size distributions if the parameters are effectively controlled. The organization of these metallic nanostructures in small clusters or lattices comprising thousands of metallic nanostructures offers another degree of freedom in controlling and tuning their properties (Klinkova, 2013).

A particularly interesting metallic nanoparticle synthesis is the so-called yolk–shell nanostructures, which are also known as A@B structures, containing a hollow space between a core material A and an outer shell B. By tuning the shape, size and composition of the inner and outer layers in both the core and the hollow shell, new properties can appear, which can be used in plenty of novel applications, such as nanoreactors, drug delivery, nanocatalysis, lithium-ion batteries and SERS (Londono-Calderon, 2016).

In the work by (Londono-Calderon, 2016), Au@AgAu yolk–shell cuboctahedra nanoparticles are formed by a galvanic replacement in a seed-mediated method. In their method, Au seeds are used to form Au@Ag core–shell nanocubes, which serve as a template material for the deposition of an external Au layer. This synthesis yields cuboctahedra nanoparticles with smooth inner and outer Au/Ag surfaces (Londono-Calderon, 2016).

Figure 5.1 shows the galvanic mechanism for the formation of Au@AgAu yolk–shell structures, which involves a growing/dissolving process to form the hollow space nanostructure. In the first step, single-crystal Au seeds are used for the formation of Au@Ag core–shell nanocubes, which serve as a template material for the deposition of an external Au layer (step 2). In step 3, the addition of Au ions triggers the galvanic replacement of Ag, which serves as a sacrificial metallic layer. The residual Ag helps keep the Au core in the middle of the hollow structure, as marked by the red arrow in Fig. 5.1 (Londono-Calderon, 2016).

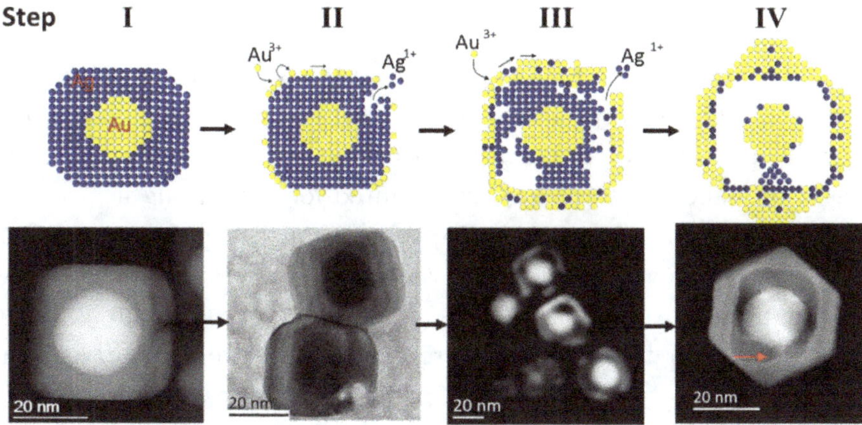

Fig. 5.1. Fabrication steps for Au@AgAu yolk–shell cuboctahedra nanoparticles. From (Londono-Calderon, 2016).

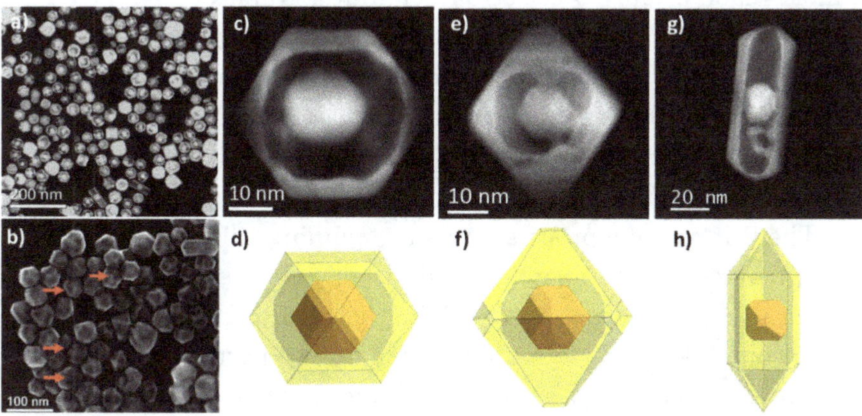

Fig. 5.2. TEM images of Au@AgAu yolk–shell structures with their respective geometrical model. From (Londono-Calderon, 2016).

Figure 5.2 shows transmission electron microscopy (TEM) and scanning electron microscopy (SEM) images of yolk–shell structures synthesized by (Londono-Calderon, 2016), in which 70% corresponded to cuboctahedra, while the rest was a combination of truncated octahedrons and nanorods.

5.2　Lithography

Top-down techniques do not usually lead to individual nanoparticles; however, they can produce bulk nanostructures. If a particular metallic nanostructured pattern is needed, then a top-down approach based on nanolithography is better suited for fabricating it.

Focused ion beam and electron beam lithography are two maskless top-down technologies that are widely used to fabricate nanostructures of different sizes or shapes very accurately on a substrate. Both these methods are expensive, have low throughput and are used for proof-of-concept or research devices. For mass production, other techniques such as soft and nanoimprint lithography are used. In these techniques, a large number of nanostructures can be fabricated on a substrate at a low cost with high accuracy using a mold made usually by electron beam lithography.

Lithography is the cornerstone of modern integrated circuit (IC) manufacturing. The ability to print patterns with submicron features and to position those patterns on a silicon substrate with better than 0.1 μm. precision is what makes ICs possible. Figure 5.3 shows a schematic diagram of a basic lithography system.

The lithographic process starts by spinning a light-sensitive photoresist onto a wafer, forming a thin layer on the surface. The resist is then selectively exposed by shining light through a mask (reticle) which contains the pattern information for the particular layer being fabricated. This exposure process modifies the resist, making it more (positive resist) or less (negative resist) soluble to a developer. After the development process, the resist will remain in some areas and be removed from some other areas, resembling the mask pattern. The process of transferring the pattern to the wafer can be done by removing areas (etching), adding materials (deposition) or modifying the characteristics of the wafer (implantation or diffusion), and the pattern transfer takes place by protecting some areas with photoresist and leaving other areas exposed to these processes.

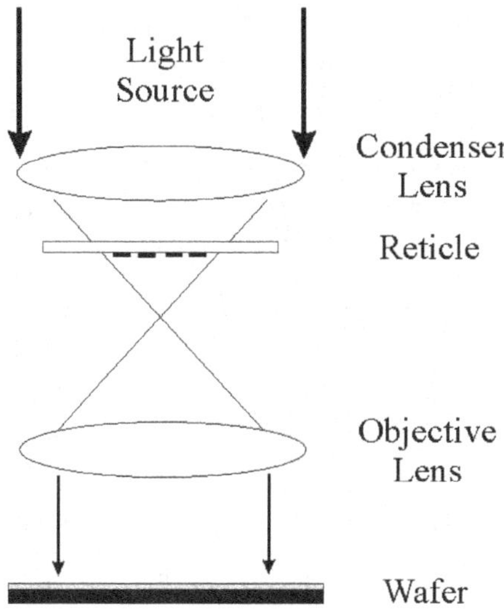

Fig. 5.3. Schematic diagram of a simple lithographic exposure system.

The photoresists used in IC fabrication normally have three components: a resin or base material, a photoactive compound (PAC) and a solvent that controls the mechanical properties, such as viscosity. In positive resists, the PAC acts as an inhibitor before exposure, slowing the rate at which the resist will dissolve when placed in a developing solution. Upon exposure to light, a chemical process occurs, by which the inhibitor becomes a sensitizer, increasing the dissolution rate of the resist. The performance of a resist is measured in terms of sensitivity and resolution. Sensitivity refers to the amount of luminous energy (usually measured in mJ/cm^2) necessary to create the chemical change described above. Resolution refers to the smallest feature that can be reproduced in a photoresist. The most popular resists are referred to as DQNs, corresponding to the photoactive compound based on diazoquinones (DQs), and the matrix material novolac (N), which dissolves easily in an aqueous solution. Solvents are added to the resin to adjust the viscosity, which is an important parameter for spin-coating the resist to the wafer.

Most of the solvent is evaporated from the resist before the exposure is completed and so plays little part in the actual photochemistry. One of the great advantages of DQN resists is that they have very good resolution since the unexposed areas are essentially unchanged by the developer because it does not penetrate the resist. Another advantage is that novolac is fairly resistant to chemical attack, making it a good mask for subsequent plasma etching. Negative photoresists swell during the development phase, broadening the linewidth. An after-develop bake will typically cause the lines to return to their original dimension, but this swelling and shrinking process often causes the lines to be distorted. As a result, negative resists are generally not suited to features less than 2.0 μm.

The most common type of optical source for photolithography is the high-pressure mercury-xenon arc lamp. Arc lamps are the brightest incoherent sources available; they emit light from a compact region a few millimeters in diameter and have total emissions from about 100 to 2000 W. A large fraction of the total power emerges as infrared and visible light energy, which must be removed from the optical path using multilayer dielectric filters. The useful portion of the spectrum consists of several bright emission lines in the near ultraviolet range and a continuous emission spectrum in the deep ultraviolet range. Because of their optical dispersion, refractive lithographic lenses can use only a single emission line, either the g line at 425.83 nm, the h line at 404.65 nm, or the i line at 365.48 nm. Each of these lines contains less than 2% of the total power of the arc lamp.

Because of diffraction effects, there is a resolution limit in lithographic projection systems, given by Rayleigh's criteria:

$$D = k_1 \frac{\lambda}{NA}, \tag{5.1}$$

where D is the minimum dimension that can be printed, λ is the exposure wavelength, and NA is the numerical aperture of the optical system. The proportionality constant k_1 is a dimensionless number in an approximate range from 0.6 to 0.8. The resolution of optical lithography using mercury arc lamps is about 0.5 μm.

5.3   E-beam Lithography

Electron beam lithography (EBL) is a specialized technique for creating extremely fine patterns. It is also used to generate masks for optical lithography (Sheats, 1998).

Derived from the early SEMs, the EBL technique consists of scanning a beam of electrons across a surface covered with a resist film sensitive to those electrons, thus depositing energy in the desired pattern in the resist film. The process of forming the beam of electrons and scanning it across a surface is very similar to what happens inside a common television or cathode ray tube (CRT) display, but EBL typically has three orders of magnitude better resolution. The main attributes of the technology are as follows:

- It is capable of very high resolution.
- It is a flexible technique that can work with a variety of materials and an almost infinite number of patterns.
- It is slow, being one or more orders of magnitude slower than optical lithography.
- The machinery required is expensive and complicated.

Figure 5.4 shows a block diagram of a typical EBL tool. The column is responsible for forming and controlling the electron beam.

Underneath the column is a chamber containing a stage for moving the sample around and facilities for loading and unloading it. Associated with the chamber is a vacuum system needed to maintain an appropriate vacuum level throughout the machine and also during the load and unload cycles. A set of control electronics supplies power and signals to the various parts of the machine. Finally, the system is controlled by a computer, which may be anything from a personal computer to a mainframe. The computer handles such diverse functions as setting up an exposure job, loading and

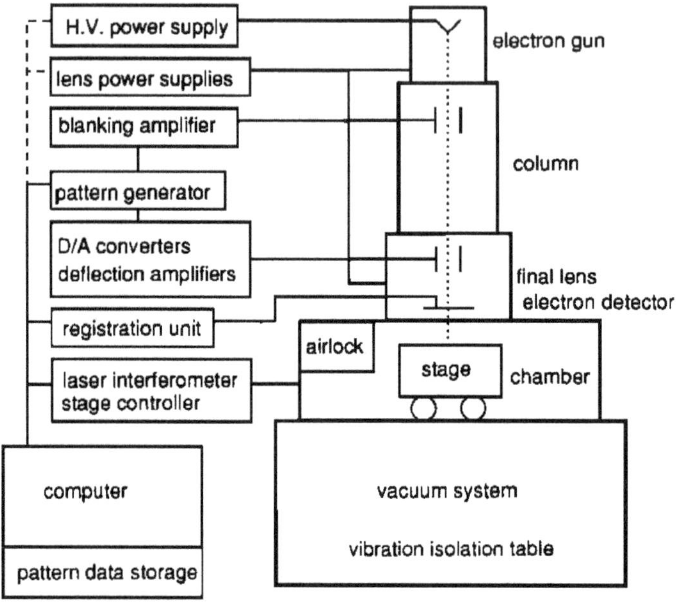

Fig. 5.4. Block diagram showing the major components of a typical electron beam lithography system. From (Rai-Choudhury, 1997).

unloading the sample, aligning and focusing the electron beam and sending pattern data to the pattern generator. The part of the computer and electronics used to handle pattern data is sometimes referred to as the data path (Rai-Choudhury, 1997).

One of the major areas of concern for EBL is pattern distortion due to proximity effects. This refers to the tendency of scattered electrons to expose nearby areas that may not be intended for exposure. There are several techniques to minimize the proximity effect; the most popular one is dose correction, where the dose is varied in such a way as to deposit the same energy density in all exposed regions of the pattern. Another way of correcting the proximity effect is shape correction, where the width of lines is decreased and the spacings between them are increased to compensate for the widening of features due to the proximity effect. The magnitude of these corrections is obtained empirically from test exposures. For this reason, this technique is generally applied only to simple, repetitive patterns; otherwise, this process would turn out to be too time-consuming. Also, the use of beam energies much greater than 20 keV (e.g., 50 keV) reduces the proximity effect because at higher energies, the electrons are scattered into a considerably larger region, giving rise to a lower concentration of scattered electrons in the pattern region.

5.4   Electron Beam Resist Processing

Electron beam resists are the recording and transfer media for EBL. The usual resists are polymers dissolved in a liquid solvent. Liquid resist is dropped onto the substrate, which is then spun at 1000–6000 rpm to form a coating. After baking out the casting solvent, electron exposure modifies the resist, leaving it either more soluble (positive) or less soluble (negative) in a developer. This pattern is transferred to the substrate either through an etching process (plasma or wet chemical) or by the "liftoff" of material. In the liftoff process, a material is evaporated from a small source onto the substrate and resist, as shown in Figure 5.5. The resist is washed away in a solvent, such as acetone. An undercut resist profile aids in the liftoff process by providing a clean separation of the material. As a rule of thumb, the thickness of the resist should be at least $3\times$ the thickness of the metallic film to get the best results using liftoff.

Polymethyl methacrylate (PMMA) was one of the first materials developed for EBL. It is the standard positive e-beam resist and remains one of the highest-resolution resists available. PMMA is usually purchased in two high-molecular-weight forms (496 K or 950 K) in a casting solvent such as chlorobenzene or anisole. PMMA is spun onto the substrate and baked at 170 to 200°C for one to two hours. Electron beam exposure breaks the polymer into fragments that are dissolved preferentially by a developer such as methyl isobutyl ketone (MIBK). MIBK alone is too strong a developer and removes some of the unexposed resist. Therefore, the developer is

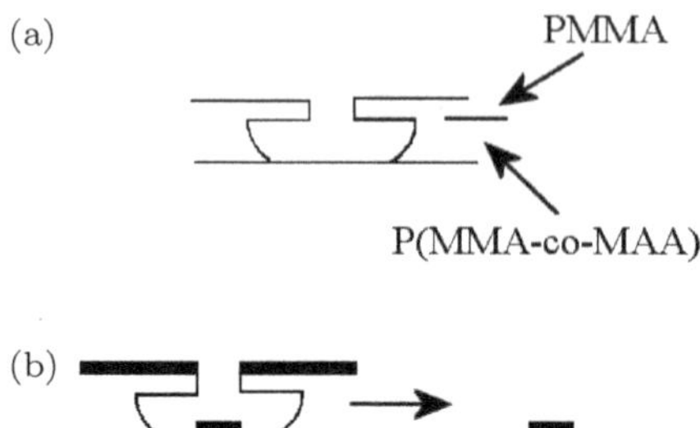

Fig. 5.5. Liftoff Process. (a) PMMA is spun on top of copolymer P(MMA-co-MAA) and developed in MIBK:IPA, giving a slight undercut. (b) Metal is evaporated and resist is removed using a liquid solvent, transferring the pattern to the substrate. From (Rai-Choudhury, 1997).

usually diluted by mixing it in a weaker developer such as isopropanol (IPA). A mixture of one part MIBK and three parts IPA produces very high contrast but low sensitivity. By making the developer stronger, say, 1:1 MIBK:IPA, the sensitivity is improved significantly with only a small loss of contrast.

The critical dose in PMMA scales with electron acceleration voltage, being roughly twice at 50 kV than at 25 kV exposures. Fortunately, electron guns are proportionally brighter at higher energies, providing twice the current in the same spot size at 50 kV. When using 50 kV electrons and 1:3 MIBK:IPA developer, the critical dose is around 350 $\frac{\mu C}{cm^2}$ (Rai-Choudhury, 1997).

When exposed to more than 10 times the optimal positive dose, PMMA will cross-link, forming a negative resist. It is simple to see this effect after exposing one spot for an extended duration (for instance, when focusing on a mark). The center of the spot will be cross-linked, leaving the resist on the substrate, while the surrounding area is exposed positively and is washed away. In its positive mode, PMMA has an intrinsic resolution of less than 10 nm. In its negative mode, the resolution is about 50 nm. By exposing PMMA (or any resist) on a thin membrane, the exposure due to secondary electrons can be greatly reduced and the process latitude is thereby increased. PMMA has poor resistance to plasma etching compared to novolac-based photoresists. Nevertheless, it has been

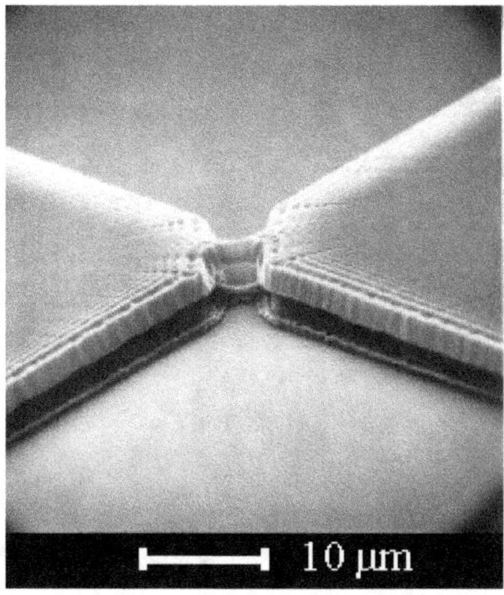

Fig. 5.6. Resist bridge pattern used to fabricate airbridge microbolometers. From (Neikirk, 1998).

used successfully as a mask for the etching of silicon nitride and silicon dioxide, with 1:1 etch selectivity (Rai-Choudhury, 1997).

A larger undercut resist profile is often needed for lifting off thicker metal layers. One of the first bilayer systems was developed by Hatzakis (Hatzakis, 1979). In this technique, a high-sensitivity copolymer of methyl methacrylate and methacrylic acid [P(MMA-MAA)] is spun on top of PMMA. A more common use of P(MMA-MAA) is as the bottom layer, with PMMA on top. In this case, the higher speed of the copolymer is traded for the higher resolution of PMMA. The undercut of this process is so large that it can be used to form free-standing bridges of PMMA (Figure 5.6).

5.5 Thin-Film Deposition Techniques

5.5.1 *Evaporation*

The metal layers for all of the early semiconductor technologies were deposited through evaporation, which has been displaced by sputtering in most silicon technologies for two reasons.

The first is the ability to cover surface topology, also called the "step coverage." Evaporated films have very poor ability to cover height discontinuities, often becoming discontinuous on the vertical walls. It is also difficult to produce well-controlled alloys through evaporation. In some cases, the poor step coverage of evaporation can be used to our advantage. Rather than depositing and etching metal layers, the film is deposited on top of a patterned photoresist layer. The films naturally tend to break at the edges of the resist so that when the resist is subsequently dissolved, the layer on top of the resist is easily lifted off (Figure 5.5) (Campbell, 1996).

In an evaporator, the wafers are loaded into a high vacuum chamber that is commonly pumped with either a diffusion pump or a cryopump. Diffusion pumped systems commonly have a cold trap to prevent the backstreaming of pump oil vapors into the chamber. The charge or material to be deposited is loaded into a heater container called the "crucible." It can be heated simply by means of an embedded resistance heater and an external power supply or by using an electron-beam gun. As the material in the crucible becomes hot, the charge gives off a vapor. Since the pressure in the chamber is much less than 1 mtorr, the atoms of the vapor travel across the chamber in a straight line until they strike a surface where they accumulate as a film.

Evaporation systems may contain several crucibles to allow for the deposition of multiple layers without breaking vacuum. To help start and stop the deposition, mechanical shutters are used in front of the crucibles, and crystal monitors are used to control the thickness of the deposited metal film.

5.5.2 *Sputtering*

Sputtering is the primary alternative to evaporation for metal film deposition in microelectronics fabrication. First discovered in 1852, sputtering was developed as a thin-film deposition technique by Langmuir in the 1920s. It has better step coverage than evaporation, induces far less radiation damage than electron beam evaporation and is much better at producing layers of compound materials and alloys. In the case of microbolometers, sputtering provides better contact between the bolometric sensor and the antenna. This is important since it has been shown that bad contacts will affect the responsivity of the microbolometer (Codreanu, 2003) and will also increase its $1/f$ noise level (Motchenbacher, 1973).

A simple sputtering system consists of a parallel-plate plasma reactor in a vacuum chamber, where a high density of ions strikes a target containing the material to be deposited. Atoms of this material are ejected and collected by the substrates that are to be coated with that material. In sputtering, the target material (not the substrate wafers) must be placed on the electrode with the maximum ion flux. To collect as many of these ejected atoms as possible, the cathode and anode in a typical sputtering system are closely spaced, often less than 10 cm. An inert gas is normally used to generate the plasma. The gas pressure in the chamber is held at about 0.1 torr. This results in a mean free path in the order of hundreds of microns.

Due to the physical nature of the process, sputtering can be used for depositing a wide variety of materials. In the case of elemental metals, simple DC sputtering is usually favored. When depositing insulating materials, such as SiO_2, an RF plasma must be used. If the target material is an alloy or compound, the stoichiometry of the deposited material may be slightly different from that of the target material (Elwenspoek, 2001).

5.5.3 *Chemical Vapor Deposition (CVD)*

Evaporation and sputtering are two types of "physical vapor deposition," where physical methods are used to produce the constituent atoms which pass through a low-pressure gas phase and then condense on the substrate. In the case of CVD, reactant gases are introduced into the deposition chamber, and chemical reactions between them on the substrate surface are used to produce the film.

CVD has historically been used in the IC industry, mainly for silicon and dielectric deposition, primarily due to its high-quality films and good step coverage.

The materials usually deposited using CVD are silicon in the polycrystalline form (polysilicon), silicon nitride and phosphor silicate glass (PSG). Polysilicon has properties comparable to single crystalline silicon, and silicon nitride is a very hard, chemically inert and strong, but brittle material with a small thermal conductivity. PSG is mainly used as a sacrificial layer for silicon micromachining.

There are three types of CVD: atmospheric-pressure CVD (APCVD), low-pressure CVD (LPCVD) and plasma-enhanced CVD (PECVD). LPCVD has better step coverage and the mechanical and chemical quality of the film (in terms of impurities, pinholes and density) is much better than in the case of APCVD. Films that will be part of mechanical microstructures should be free of internal stresses, or else bending and buckling will occur. The stress of a film grown on top of a substrate is indicated with respect to the underlying material. An expanding layer is then said to be under compressive stress, while a contractive layer is under tensile stress (Plummer, 2000). Bending and buckling will occur if the film is under compressive stress. Low-stress nitride films can be deposited in an LPCVD reactor at 835°C. PECVD is used when the deposition needs to be done at low substrate temperatures (~300°C). This deposition method has good step coverage, but the films suffer from pinholes and a high hydrogen concentration, making them less suitable for mechanical microstructures.

5.6　Etching

After thin films are deposited on the wafer surface, they can be selectively removed by etching to leave the desired pattern on the wafer surface. In addition to deposited films, parts of the silicon substrate itself may be etched, such as in creating trenches in isolation structures. The masking layer may be a photoresist, or it may be another thin film, such as silicon dioxide or silicon nitride. Oxide or nitride masks stand up better than photoresist to etching conditions and are often called hard masks. But they themselves must be selectively etched, usually using a lithographically defined photoresist as the masking layer. The etching of a thin film is usually done until a different layer (known as "etch stop") is reached underneath.

Etching can be done in either a "wet" or "dry" environment. Wet etching involves the use of liquid etchants. The wafers are immersed in the etchant solution, and the exposed material is etched mostly through chemical processes. Dry etching involves the use of gas-phase etchants in a plasma. Here, the etching usually takes place through a combination of chemical and physical processes. Because a plasma is involved, dry etching is usually called "plasma etching" (Lyshevski, 2002). Both methods can be either isotropic, i.e., provide the same etch rate in all directions, or anisotropic, i.e., provide different etch rates in different directions (Figure 5.7). The important criteria for selecting a particular etching process are the material etch rate, the selectivity to the material to be etched versus other materials, and the isotropy/anisotropy of the etching process.

Wet etching provides a good etch selectivity and is usually isotropic, with the exception of anisotropic silicon wet etch using potassium hydroxide (KOH). Dry etching is often anisotropic, resulting in a better pattern transfer, as mask underetching is avoided (Figure 5.7). Reactive ion etching (RIE) is a common form of dry etching where reactive ions are generated in a plasma and accelerated toward the surface to be etched. This process is anisotropic but has low etch selectivity.

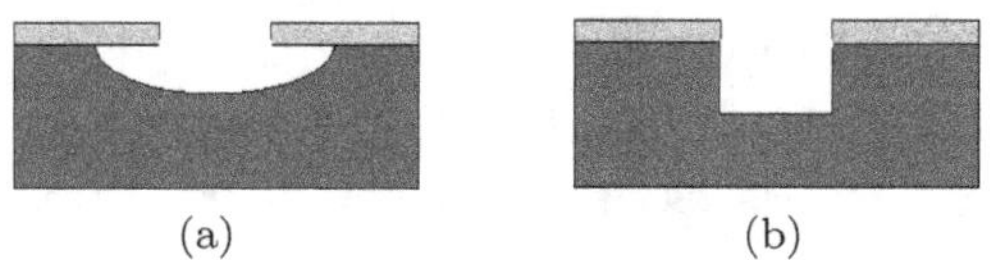

Fig. 5.7. Schematic of: (a) Isotropic and (b) anisotropic thin-film etching.

5.7 Silicon Micromachining

Silicon micromachining is a process used to fabricate static and movable 3D microstructures, such as bridges, cantilevers and membranes, on silicon substrates. There are two types of silicon micromachining: bulk micromachining and surface micromachining. In the case of bulk micromachining, wet-etch and dry-etch techniques are used to remove parts of the silicon substrate and form the microstructure, whereas in the case of surface micromachining, the microstructure is made of thin-film layers which are deposited on top of the substrate and selectively removed in a defined sequence.

Surface micromachining has become the major fabrication technology of microscale structures because it uses standard CMOS fabrication processes and facilities. The most commonly used surface micromachining process is sacrificial-layer etching. In this process, a microstructure is released by removing a sacrificial thin-film material which was previously deposited underneath the microstructure (Figure 5.8).

Usually, the sacrificial layer is made out of silicon dioxide (SiO_2), phosphorous-doped silicon dioxide (PSG) or silicon nitride (Si_3N_4), and the structural layers are then typically formed with polysilicon, metals or alloys. There are three key challenges in fabricating microstructures using surface micromachining: control and minimization of stress and stress gradient in the structural layer to avoid bending or buckling of the released microstructure; high selectivity of the sacrificial layer etchant to structural layers and silicon substrate; and avoidance of stiction of the suspended microstructure to the substrate (Lyshevski, 2002).

Stiction is the tendency of the suspended structures to collapse due to the surface tension of liquids during evaporation. The liquid forms a droplet during drying between the microstructure and the substrate, which generates an underpressure that will make the structure collapse if it is not stiff enough (Figure 5.9).

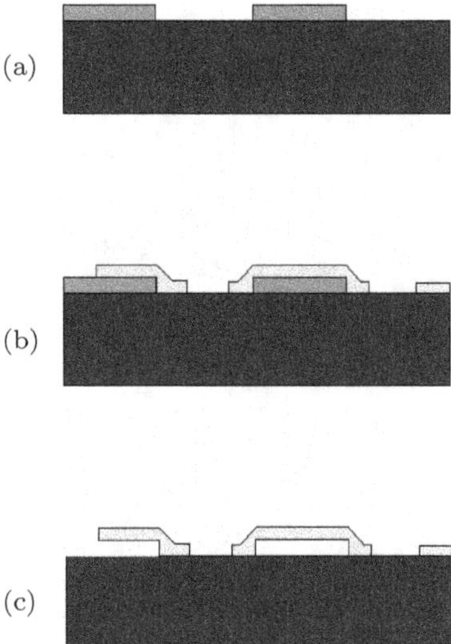

Fig. 5.8. Surface micromachining fabrication process: (a) deposition and patterning of the sacrificial layer; (b) deposition and patterning of the structural layer; (c) release etch.

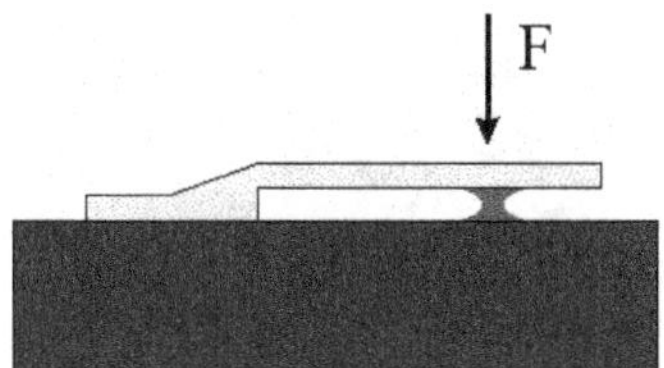

Fig. 5.9. Capillary force during sacrificial layer etch.

Rinsing procedures such as critical-point drying or freeze drying can be used to avoid stiction. Critical-point drying works by exchanging the rinsing liquid, after etching the sacrificial layer, with liquid carbon dioxide, which is then removed in its supercritical state avoiding the liquid–gas phase transition. An alternative procedure to avoid stiction is to introduce some roughness to the structure or use an anti-stiction coating.

Part 4

Characterization and Modeling

In this part, we delve into the critical processes of characterization and modeling of metallic nanostructures, essential for understanding their properties and optimizing their applications. Accurate characterization is fundamental to the advancement of nanotechnology, and we will explore key techniques such as scanning electron microscopy and transmission electron microscopy, which provide invaluable insights into the morphology and structural details of nanostructures. Additionally, we will discuss various spectroscopy techniques that reveal the chemical composition and electronic properties, enhancing our understanding of these materials at the nanoscale. Transitioning from characterization to modeling, we will examine the various numerical techniques employed to simulate the behavior of metallic nanostructures. Emphasis will be placed on the finite element method, a powerful tool for analyzing complex geometries and physical phenomena. We will also highlight the capabilities of COMSOL Multiphysics, a commercial software that facilitates the modeling of multiphysics interactions, allowing researchers to predict performance and optimize designs effectively.

Chapter 6

Characterization

Metallic nanostructures can be characterized by their morphology and physical properties, which include optical, mechanical and electrical properties, among others (Murty, 2013). Morphology of metallic nanostructures can be investigated using electron microscopy, while other properties require specialized equipment for their characterization, such as optical spectroscopy and near-field scanning optical microscopy for optical properties and nanoindentation to characterize mechanical properties.

In this chapter, the instruments used to morphologically and optically characterize metallic nanostructures are presented.

6.1 Scanning Electron Microscopy (SEM)

Scanning electron microscopy (SEM) is a fundamental technique for morphological characterization. It can be used to characterize nanostructures of even a few nanometers. In addition to shape, size and surface topography, SEM can provide information on chemical composition, crystal orientation and internal stress distribution through energy dispersive X-ray spectroscopy (EDS) (Murty, 2013).

The high resolving power of electron microscopes is due to the de Broglie wavelength, which is the wavelength associated with particles, such as the electron, and the resolution limit of optical systems, which is proportional to the wavelength of the illuminating source.

The theoretical limit for optical systems is due to diffraction. The resolving power of an instrument can be derived from Rayleigh's criterion, which states that two images are resolvable when the center of the diffraction pattern of one is directly over the first minimum of the diffraction pattern of the other. Using this criterion, the resolving power of an instrument is given by

$$R = 1.22 \frac{\lambda}{NA}, \tag{6.1a}$$

where λ is the wavelength of the illumination source and NA is the numerical aperture, which is a dimensionless number used to describe the acceptance angle of the light that enters or exits a lens. It is defined by $NA = n\sin(\theta)$, where θ is the half-angle of the cone of light that can enter or exit the lens.

The de Broglie wavelength of an electron can be calculated from $\lambda = h/p = h/\sqrt{2mE}$, where h is the Planck constant, p is the momentum of the electron, m the mass of the electron, and E the energy of the electron. For an electron beam with an energy of 25 KeV, the equivalent de Broglie wavelength is $\lambda = 7.8 \times 10^{-12} m$, which is two orders of magnitude smaller than the size of an atom. Considering that the illumination used in optical microscopy is in the $400 \times 10^{-9} m - 700 \times 10^{-9} m$ wavelength range, and electron microscopes can generate electron beams in the range of a few hundred eV to 50 KeV, then the optical resolution of an electron beam would be five orders of magnitude better than that of an optical microscope.

A typical SEM consists of an electron gun that emits electrons focused on a beam with a very small spot size of $\sim 5nm$. Electrons are accelerated at a particular energy, which will provide the required resolution and will not burn the sample. The electron beam is then scanned along a pattern of parallel lines. Focusing and deflecting the electron beam for the raster scan is performed by deflection coils. Detectors collect the electrons scattered from the sample and are

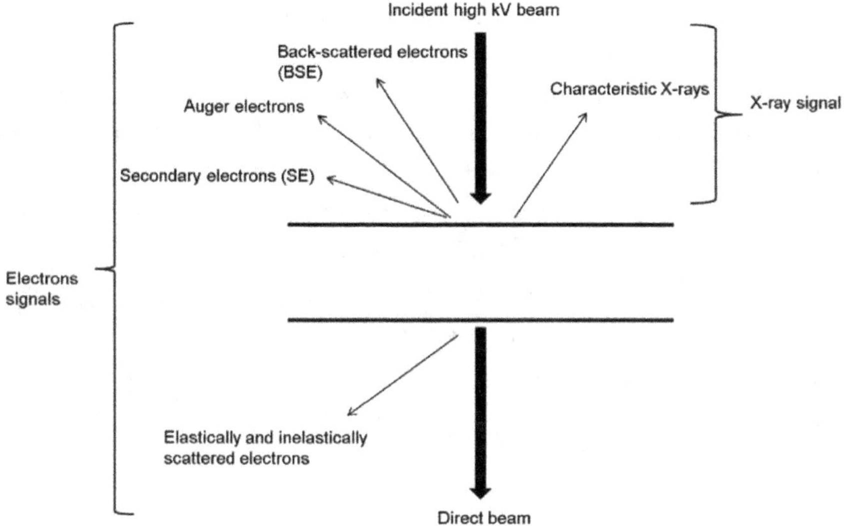

Fig. 6.1. Types of signals produced by high-energy electrons hitting a sample. Reprinted from (Tang, 2017).

used to generate an image or to analyze the sample surface (Murty, 2013).

There are different SEM techniques based on the type of electrons detected and imaged, mainly: secondary electron imaging, backscattered electron imaging and elemental X-ray detection (Murty, 2013). The type of electrons depends on the type of interaction between the high-energy primary that comes from the electron beam and the atoms of the sample (Figure 6.1).

If the primary electron undergoes inelastic scattering, then part of its energy is transferred to the electrons of the sample. If the energy transfer is large enough, the sample will emit electrons. If the emitted electron has an energy of less than 50 eV, it is known as an *electron*. Imaging secondary electrons are highly sensitive to topographic variations.

If the primary electrons undergo elastic scattering, then the scattered electrons have the same energy as the incident electrons, and these electrons are known as *backscattered electrons*. The probability of backscattering increases with the atomic number of the sample material, making the imaging of backscattered electrons useful in providing atomic number contrast in addition to topographical contrast.

Another type of interaction of electrons takes place when the primary electron collides with an electron of the sample and ejects a core electron from the atom in the sample. This will generate an X-ray photon or an Auger electron, both of which can be used for chemical characterization through EDS or Auger spectroscopy. Figure 6.2 shows the three characterization results of nanoparticles obtained using different SEM modes.

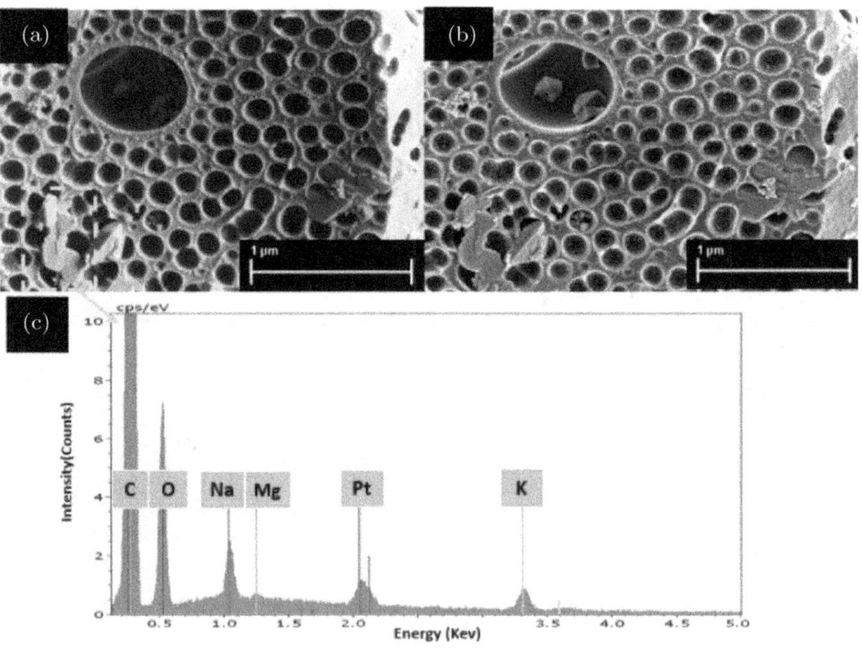

Fig. 6.2. Examples of electron microscopies in (a) secondary emission mode, (b) backscatter image mode and (c) X-ray spectroscopy mode. Reprinted from (Jalalabadi, 2020).

A fundamental part of electron microscopes is the electron gun. The electron gun can be thermal, where a filament made of tungsten or LaB_6 is heated via Joule heating until it emits electrons, which travel through the vacuum chamber at the acceleration voltage. The other type of electron gun is the field-emission (FE) gun, which uses the tunneling or Schottky effect to emit electrons from the tip. It uses a very high electric field $(\sim 10^6 V/cm)$ applied at the apex of the tip for electron emission. FE-type guns have very high brightness, which is about three orders of magnitude higher than conventional "thermal-emission" (TE) guns (Kogure, 2013).

The FE gun using the tunneling effect does not heat the tip during emission, so it is called a cold FE gun, while the other gun using the Schottky effect heats the tip. The cold FE gun is brighter and has a larger source size and energy spread, while the Schottky FE gun can emit a higher and more stable current, which is desired for X-ray chemical analysis, electron backscatter diffraction and electron beam lithography.

6.2 Transmission Electron Microscopy (TEM)

Transmission electron microscopy (TEM) is a high-resolution imaging technique in which a beam of electrons travels through a thin sample to produce an image (Figure 6.3). TEM can provide microstructural, crustal structure as well as micro-chemical information with a high spatial resolution.

In TEM, an electron beam with an energy of 100 keV or more (up to 3 MeV) passes through a thin specimen (less than 200 nm) under a high vacuum. The electron beam is then affected by the specimen, and the resulting image is projected on a fluorescent screen. The thickness in nanometers of the samples should be less than two times the applied accelerating voltage in kV.

The contrast in TEM images is given by the scattering or absorption of electrons by the sample, which absorbs or deflects electrons in directions different from the direction of the beam.

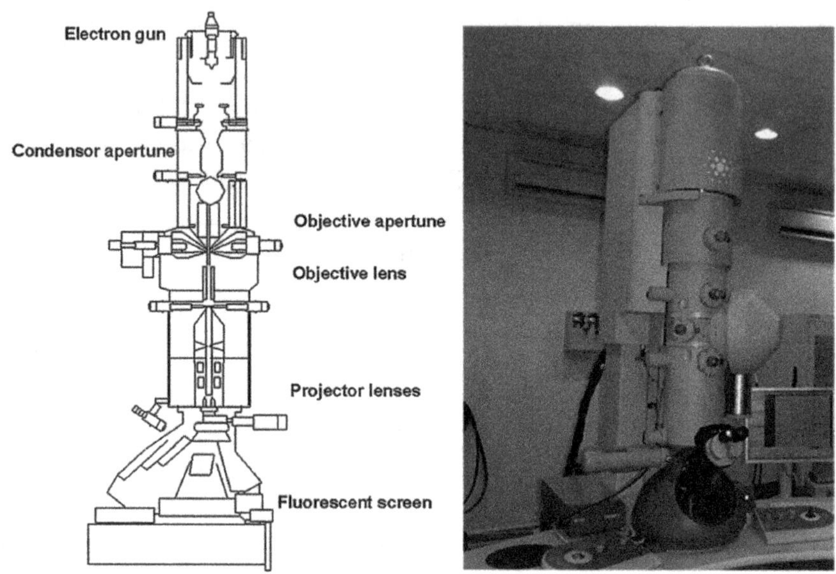

Fig. 6.3. Schematic diagram of a transmission electron microscope (TEM). Reprinted from (Tang, 2017).

Absorption and scattering depend on the atomic number of the scattering atom (i.e., the density of the material) and the local thickness of the sample: the higher the atomic number and the sample thickness, the greater the number of electrons scattered to larger angles. These scattered electrons result in dark spots, where the higher atomic number or thicker materials are located. For specimens with very low differences in density and object thickness, the contrast can be enhanced by staining techniques. These techniques involve the incorporation of heavier atoms, such as osmium or ruthenium in the sample.

A way to improve the resolution of TEM images is to increase the accelerating voltage, which decreases the equivalent de Broglie wavelength of the electrons. High-resolution electron microscopy can be achieved by using aberration correctors to correct spherical and chromatic aberrations. With these correction aberration techniques, resolutions of less than 0.05 nm can be achieved.

Electron holography and energy filtering electron microscopy have been developed to improve TEM images. In electron holography, an interference pattern formed by the electron wave is recorded and represented in terms of amplitude and phase. The phase can give information about the electric and magnetic fields at the nanoscale. Energy filtering is used to produce an enhanced contrast image by adjusting a slit so that only electrons with a certain energy can pass through. This prevents inelastic scattering from contributing to the image, which enhances the contrast of the resulting TEM image.

6.3 Optical Spectroscopy

When light interacts with matter, generally two fundamental properties of light change: intensity and polarization. These two properties can be used to determine the refractive index and extinction coefficient of materials, which are the optical properties of materials used to calculate and numerically simulate optical processes.

The refractive index and extinction coefficient of materials can be obtained experimentally by monitoring changes in both intensity and polarization when light interacts with the sample. The quantities that can be measured experimentally due to the interaction of light with a sample are transmittance, reflectance and ellipsometry.

Optical transmittance is defined as the ratio of the transmitted I_t and incident light I_0, and it is usually measured at normal incidence. Reflectance is defined as the ratio of the reflected I_r and incident light I_0. When measured at aoblique incidence, the reflectance depends on the polarization and angle of incidence of the incoming light. When defining reflectance, two linear polarizations are considered: the p polarization, which is parallel to the plane of incidence, and the s polarization, which is perpendicular.

Transmittance and reflectance are real quantities that range from 0 to 1, and they measure the ratio of light intensities, while the transmission and reflection coefficients are complex numbers providing information about not only the change in amplitude but also the phase change of the wave.

The polarization state of a reflected wave generally takes an elliptical form, so the experimental measurement of the state of polarization of the reflected or transmitted wave with respect to the polarization state of the incident wave is called ellipsometry.

6.4 Scanning Near-Field Optical Microscopy (SNOM)

Scanning near-field optical microscopy (SNOM) is an optical technique that can resolve subwavelength structures by imaging evanescent waves generated by the interaction of light, the sample and a sharply pointed probe. In the case of conventional microscopy, the resolution is limited by diffraction, which, in the far field, is given by the equation $D_{limit} = 1.22 \times \lambda \times f/\#$. This far-field diffraction limit can be overcome by imaging evanescent waves in the near field, which is located at distances shorter than the wavelength. By doing this, sub-nanometer structures can be resolved using a visible optical source. SNOM consists of probing evanescent waves generated at the sample by the interaction of light and a sharply pointed probe. Evanescent waves decay rapidly with distance, so the probe needs to be in close proximity to the sample. Two types of probes are used for SNOM: one that uses an aperture to confine and sample electromagnetic fields, and a second one that uses an antenna to couple locally enhanced near fields to propagating fields and vice versa.

Most SNOM implementations use an atomic force microscope (AFM) or a scanning tunneling microscope to keep the tip–sample distance constant.

A diagram of an s-polarized SNOM set-up operating in the far infrared is shown in Fig. 6.4. The operational principle is based on the Michelson interferometer, in which a collimated beam from a CO_2 laser (operating at $\lambda = 10.6$ μm) and linearly polarized along the z axis is divided into two by a beam splitter (BS): the first toward

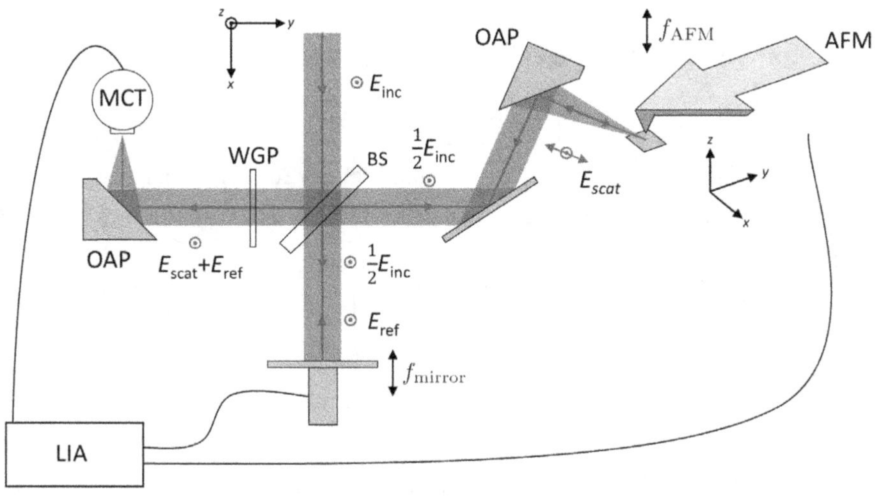

Fig. 6.4. Experimental near-field s-SNOM measurements. AFM topography of the classic bowtie and discrete bowtie are shown in (a) and (f), respectively. Modulus of the near field is shown in (b), (d), (g) and (i), and phase is shown in (c), (e), (h) and (j). Scans (b) and (c) correspond to the classical bowtie when illuminated with vertically polarized light, while (d) and (e) when illuminated with linearly polarized light at $-45°$. Scans (g) and (h) correspond to the discrete bowtie when illuminated with vertically polarized light, while (i) and (j) when illuminated with linearly polarized light at $-45°$.

the reference beam, and the second toward the sample. The reference mirror oscillates axially with a frequency of $f_{\mathrm{mirror}} = 300$ Hz and an amplitude of $\Delta L = 1.13$ μm. The second beam is reflected at the BS and then focused by an off-axis parabolic mirror (OAP) into the sample at a 60° incidence angle. The sample under test is placed in an AFM operating in the tapping mode, so the cantilever tip oscillates at frequency $f_{\mathrm{AFM}} = 250$–270 kHz. The incident field is p-polarized, so it polarizes the tip, inducing surface charges in the sample. This leads to near-field interaction between the tip and the sample. The scattered field is reflected by the same OAP mirror back to the BS. The reference and scattered fields are added, and a wire grid polarizer is used to suppress background scattering coming from the tip shaft.

Both fields are focused by a OAP mirror into a liquid-nitrogen-cooled mercury cadmium telluride (MCT) detector. A lock-in amplifier (LIA) measures the detector signal at the first and second sidebands of the second harmonic of the tip oscillation, $S_{2,1}$ and $S_{2,2}$, respectively. As outlined by (Moreno, 2017), the near-field is $E_z(x,y) = |E_z(x,y)|\exp[i\phi(x,y)]$, where the magnitude and phase are calculated as

$$|E_z(x,y)| = \sqrt{[S_{2,1}(x,y)/J_1(\alpha)]^2 + [S_{2,2}(x,y)/J_2(\alpha)]^2}, \qquad (6.2)$$

$$\phi(x,y) = \Psi_G - \Phi(x,y), \qquad (6.3)$$

where

$$\Phi(x,y) = \arctan\left[\frac{S_{2,1}(x,y)/J_1(\alpha)}{S_{2,2}(x,y)/J_2(\alpha)}\right]. \qquad (6.4)$$

J_m is the Bessel function of the first kind and mth order, $\alpha = 4\pi\Delta L/\lambda$ is the modulation depth, and Ψ_G is the so-called global phase, which is related to the mean position of the vibrating reference mirror.

Chapter 7

Numerical Modeling

The design and analysis of metallic nanostructures require an accurate near-field, which can be achieved using computational methods. The complexity of the electromagnetic field in the presence of arbitrarily-shaped nanoparticles is such that Maxwell's equations must be solved using numerical methods and rules out purely analytic methods, such as Mie theory, which is restricted to simple shapes like spheres.

Numerical methods can be classified as time domain and frequency domain methods. Among time domain methods are the finite difference time domain (FDTD) and the finite element method (FEM). These methods allow better representation of arbitrarily shaped structures. The FEM relies on an unstructured mesh, which allows a more accurate representation of curved surfaces, whereas in FDTD the discretization is performed into rectangular-shaped elements in two dimensions or cuboid elements in three dimensions (Karamehmedovic, 2011). Cuboid elements, where the electric fields are located on the edges of the cuboid and the magnetic fields are normal to the faces, are called Yee cells and are the fundamental elements of most FDTD methods.

Unstructured meshes also increase efficiency and avoid unphysical behavior due to the staircase approximation in curved surfaces in FDTD. The representation of the material properties, especially frequency-dependent permittivity as in the case of metals, also depends on the method. Time domain methods usually require an approximation of the permittivity with analytic models, such as the Drude–Lorentz model (Karamehmedovic, 2011).

7.1 Boundary Element Method (BEM)

In this method, the electromagnetic field scattered by nanoparticles is expressed in terms of boundary charges and currents. These charges and currents are obtained using the boundary conditions of the parallel and perpendicular components of the electric and magnetic fields across the nanoparticle's interface. These boundary conditions lead to a system of equations that can be solved numerically using linear-algebra techniques.

7.2 Discrete-Dipole Approximation (DDA)

In the discrete dipole approximation (DDA), the scattering is calculated by approximating the nanoparticle by a finite point array. In response to a local electric field, each point acquires a dipole moment that generates the scattered field. This method is flexible regarding the geometry of the nanoparticle, and the only limitation is that the inter-dipole separation has to be small compared to the size of the nanoparticle and the wavelength.

7.3 Finite Difference Time Domain (FDTD)

The FDTD method solves Maxwell's equations by discretizing in space and time their differential forms. Then, the equations are solved directly for every spatial grid and for every time step. This method gives the electric field as a function of time and position $E(r, t)$.

FDTD was introduced by Yee in 1966 as an alternative approach to the frequency-domain integral equation solution of Maxwell's equations. In this new approach, the direct time/domain solutions of Maxwell's differential equations are obtained on spatial lattices.

FDTD as well as other space-grid time-domain techniques are based on volumetric sampling of the unknown electric and magnetic fields in and surrounding the structure of interest over a period of time. The sampling in space is performed at a subwavelength resolution, which is set by the user to properly sample the highest near-field spatial frequencies. Typically, 10–20 samples per wavelength are ideal to get accurate near-field results. The sampling time is selected to ensure numerical stability of the algorithm.

Space-grid time-domain techniques such as FDTD are marching-in-time procedures that simulate the continuous actual electromagnetic waves propagating in a finite spatial region. Time-stepping continues as the numerical wave analogs propagate in the space lattice until it connects to the boundary conditions set by the physics of the modeled region. For simulations where the modeled region must extend to infinity, absorbing boundary conditions (ABCs) are employed at the outer lattice truncation planes, which simulate all outgoing wave analogs to exit the region with negligible reflection.

There are two types of ABCs for FDTD formulations: analytical boundary conditions imposed on the electromagnetic field at the outermost planes of the space lattice and the perfectly matched layer (PML) which is absorbing media adjacent to the outer planes of the space lattice.

Phenomena such as induction of surface currents, scattering and multiple scattering, aperture penetration and cavity excitation are modeled time-step by time-step by the action of the numerical analog to the curl equations. Self-consistency of these modeled phenomena is generally ensured if their spatial and temporal variations are well resolved by the space and time sampling process. The goal is to provide a self-consistent model of the mutual coupling of all of the electrically small volume cells constituting the structure and its near field, even if the structure spans tens of wavelengths in three dimensions and there are hundreds of millions of space cells.

Time-stepping is continued until the desired late-time pulse response is observed at the field points of interest. For linear wave interaction problems, the sinusoidal response at these field points can be obtained over a wide band of frequencies by discrete Fourier transformation of the computed field versus time waveforms at these points.

7.4   Finite Element Method (FEM)

The FEM is a numerical technique used to solve boundary value problems (BVPs) governed by a set of differential equations. This method divides the simulation domain into smaller subdomains called *finite elements*. The variables to be solved are interpolated inside the finite elements, starting with the initial values at the boundaries, which are set by the boundary conditions. The interpolation or shape functions must be a complete set of polynomials. The accuracy of the solution depends on the size and number of the subdomains created and the order of the interpolation polynomials, which may be linear, quadratic or of higher order. A system of equations is generated for each finite element using the interpolation polynomials, and the solution is obtained after solving a system of linear equations. The higher the number of finite elements, the larger the matrix needed to solve this system of equations; for this reason, the FEM usually requires a large amount of computer memory.

To generate the linear system of equations, the differential equations to be solved and their associated boundary conditions must first be converted into an integro-differential formulation by either minimizing a functional or using a weighted residual method, such as the Galerkin approach. This formulation is applied to a single element to obtain its element equations. The assembly of all elements results in a global matrix system that represents the entire domain of the BVP.

In the FEM, the Dirichlet boundary conditions are used where the value of the variable is defined at the domain boundary; on the other hand, for the Neumann boundary conditions, a solution is assumed for the derivative of the variable.

The major steps in the application of the Galerkin method FEM for the solution of a BVP are the following (Polycarpou, 2006):

(1) Discretize the domain using finite elements.
(2) Choose proper interpolation functions (also known as *shape functions* or *basis functions*).
(3) Obtain the corresponding linear equations for a single element by first deriving the weak formulation of the differential equation subject to a set of boundary conditions.
(4) Form the global matrix system of equations through the assembly of all elements.
(5) Impose the Dirichlet boundary conditions.
(6) Solve the linear system of equations using linear algebra techniques.
(7) Postprocess the results.

7.5 Commercial Software for Numerical Modeling

Commercial and open-source software based on some of the numerical techniques explained previously are commonly used to solve Maxwell's equations. Meep and Lumerical are software packages used for electromagnetic simulation based on the FDTD method. Meep is free and open source, while Lumerical has to be licensed.

ANSYS HFSS was one of the first tools in the market. It provides 3D electromagnetic simulation and has been used extensively as a tool for the design of high-frequency electronic products, such as antennas, antenna arrays and printed circuit boards. It is based on the FEM and can be used to extract parasitic parameters (S, Y, Z), which are very popular in RF and microwave design. Ansys HFSS also allows for the visualization of 3D electromagnetic fields, near and far fields and for generating SPICE models.

CST Microwave Studio is a general purpose software that has gained a lot of popularity in the microwave field. It is based on the finite integration technique and is used for the design and optimization of electromagnetic components and systems.

Lumerical was acquired by ANSYS in 2020 as part of the ANSYS software portfolio. It is based on the FDTD algorithm and can be used for 2D and 3D structures. It can also be used for modeling optical, electrical, thermal and quantum effects at the physical level. Lumerical has tools geared toward the design and optimization of the performance of photonic integrated circuits.

COMSOL Multiphysics is a partial differential equation solver based on the FEM. Maxwell's equations, as well as any other partial differential equation, can be solved using this software package. It has the advantage that different equations can be combined to solve multiphysics problems, such as electromagnetic heating in thermoplasmonic applications.

7.6 Numerical Modeling of a Large Array of Periodic Metallic Nanostructures using COMSOL Multiphysics

COMSOL Multiphysics can be used to calculate the interaction of electromagnetic waves with metallic nanostructures, including infinite arrays of metallic nanostructures. In order to calculate a large array of periodic nanostructures and optimizing the amount of memory used in the simulations a set of periodic boundary conditions can be used in the simulations. In COMSOL Multiphysics Floquet-periodic boundary conditions can be used on four sides of the unit cell to simulate an infinite array.

The Floquet boundary conditions available in COMSOL Multiphysics arises from the Floquet-Bloch theorem used to solve systems of differential equations with periodic coefficients. The Floquet theorem is used to solve ordinary differential equations of the form:

$$\frac{d\omega(x)}{dx} = A(x)\omega(x), \tag{7.1}$$

where $\omega(x)$ is the unknown function, and $A(x)$ is a given matrix of continuous periodic functions with period r_1.

The Bloch theorem was originally introduced to represent the form of homogeneous states of Schrodinger equation with periodic potential. This theorem can be considered as a multidimensional application of the Floquet theorem.

The procedure to perform simulations with COMSOL Multiphysics starts with defining the physics to be used and the geometry of the simulation, in a 3D simulation a simulation volume needs to be defined as well as the materials that constitute the volume and its physical properties (Figure 7.1). Once the physics and the geometry has been defined, the boundary conditions are defined. Perfectly matched layers (PMLs) on top and bottom of the unit cell will absorb the energy from the source port and any higher modes generated by the periodic structure. The PMLs attenuate the

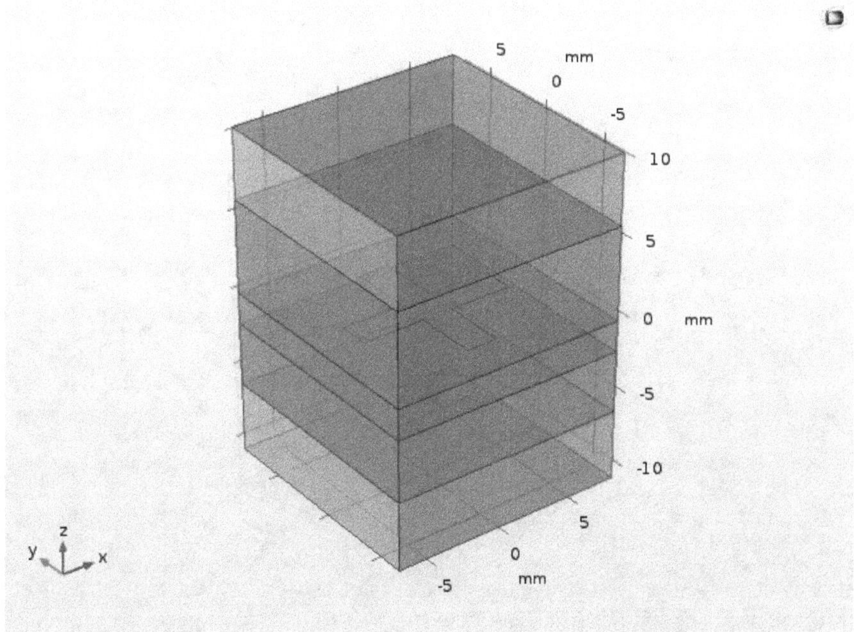

Fig. 7.1. A typical three dimensional volume used to perform 3D electromagnetic simulations with COMSOL Multiphysics.

wave as it propagates in the direction perpendicular to the PML boundary.

Port boundary conditions are used to generate the incident electromagnetic wave and to automatically determine the reflection and transmission characteristics of the periodic array.

After the simulation volume and the boundary conditions are set the volume needs to be divided into finite elements, this is done by generating a mesh for the solution of the systems of equations using the finite element method, in the case of electromagnetic problems a mesh with elements smaller than 1/20th of the incident wavelength should be used to guarantee an accurate solution (Figure 7.2). Depending if the simulation is two or three dimensional the mesh elements can be triangles for 2D and tetrahedrons for 3D simulations.

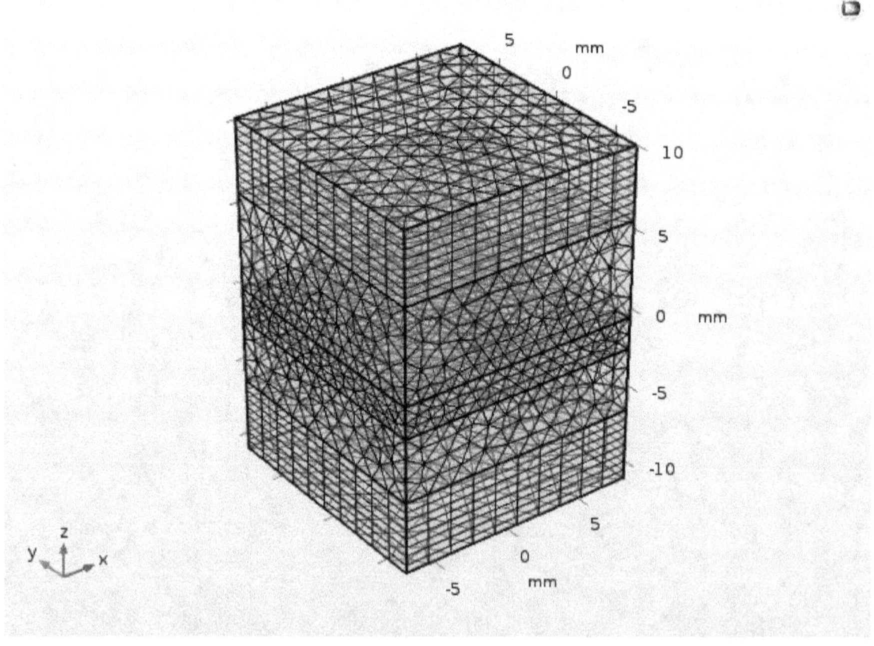

Fig. 7.2. Discretization of the 3D simulation volume by using tetrahedrons.

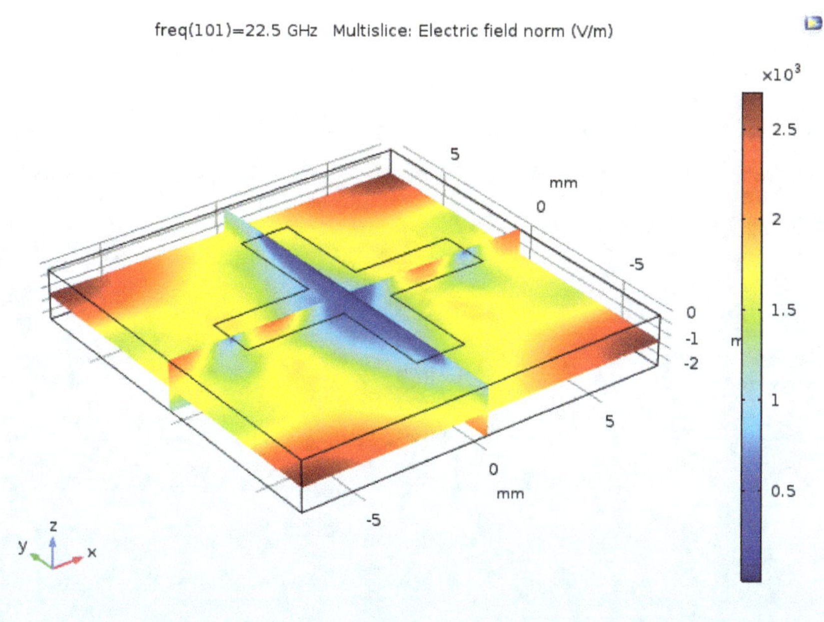

Fig. 7.3. Graph of the calculated electric field on a wave propagation simulation solved using the finite element method.

The postprocessing section of COMSOL Multiphysics allows to plot the electromagnetic variables solved, such as the electric and magnetic fields on all the simulation volume (Figure 7.3), as well as to calculate derivative variables such as the transmission and reflection coefficients.

Part 5

Applications

In this part, we explore the diverse and transformative applications of metallic nanostructures, showcasing their potential to revolutionize various fields — from plasmonic applications that harness the unique interactions between light and metallic nanoparticles to the innovative realm of thermoplasmonics. We will delve into the fascinating world of metasurfaces, which enable unprecedented control over electromagnetic waves, paving the way for advancements in optics and imaging. Additionally, we will discuss surface-enhanced Raman spectroscopy, a powerful technique that leverages metallic nanostructures to amplify signals for ultrasensitive detection, making it invaluable in chemical and biological sensing. Biosensors represent another exciting application, where metallic nanostructures enhance sensitivity and specificity, leading to breakthroughs in medical diagnostics and environmental monitoring. A special emphasis will be placed on optical antennas, which exemplify the intersection of nanotechnology and photonics, enabling efficient light manipulation and signal enhancement.

Chapter 8

Thermoplasmonics

The heating of metallic nanostructures under electromagnetic radiation has opened the path to thermoplasmonics, an interesting field with applications which include photothermal therapy, drug release, thermal-optical data storage, solar thermal energy harvesting and optoelectronic devices, among others (Herzog, 2014).

The transfer of optical to thermal energy takes place through absorption and resonant coupling into plasmon modes. Thermal analysis of optical heating of metallic nanostructures can be determined by balancing incoming optical energy with outgoing transfer via conduction, radiation and convection. Energy transfer equations can be solved through numerical computation using the finite element method or any other differential equation solving technique. Experimental analysis of thermal processes in plasmonic nanostructures can be performed using scanning thermal microscopy or thermographic phosphor techniques.

8.1 Photo-Thermoelectric Conversion

Plasmonic local heating can be used to generate a thermoelectric current for light detection or energy harvesting, with the advantage that the electromagnetic energy that generates the increase in temperature will occur only for the plasmonic wavelength of resonance, creating a frequency-selective heater or harvester.

Light-detection systems based on thermoelectric conversion using carbon nanotubes, graphene, carbon sprays and commercial thermoelectric devices coupled to metallic nanostructure plasmon resonances have been reported (Miwa, 2020). By using plasmon resonances to increase the temperature in thermoelectric devices, photodetection and electromagnetic energy harvesting from the visible to the mid-infrared (IR) region can be achieved (Miwa, 2020).

Figure 8.1 shows a thermoplasmonic photodetector made of a 40 nm thick silver film deposited through thermal evaporation with nanoholes patterned using electron beam lithography and liftoff as the plasmonic element.

Testing of a thermoplasmonic photodetector can be done using a laser at the wavelength of the plasmon resonance of the plasmonic element, and the thermoelectric response can be measured with an electrometer. For a thermoplasmonic device, like the one in Figure 8.1, with hole diameters of (192 ± 3.22) nm and a periodicity of 300 nm, the extinction peak was observed at 687 nm, indicating excitation of surface plasmon polaritons on the Ag nanohole array. The photocurrent under 690 nm was measured at 173 nA when illuminating the nanohole array and 11 nA when illuminating the Ag film where no plasmonic structures were present. This current was proportional to the optical power.

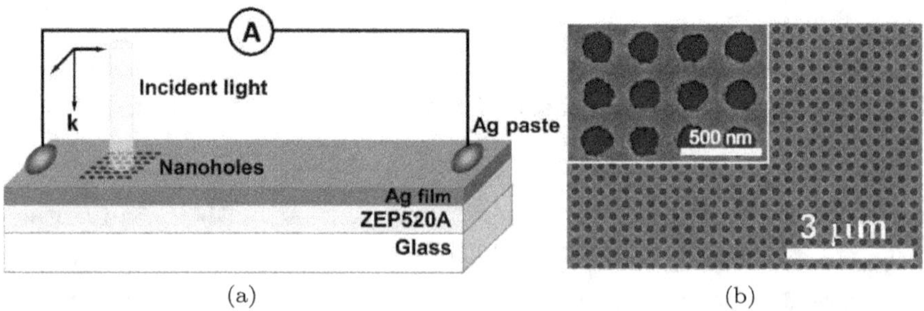

Fig. 8.1. (a) Schematic diagram of a silver thin-film thermoelectric photodetector. (b) Scanning electron micrograph of the periodic plasmonic nanohole arrays of a plasmonic thermoelectric photodetector. Taken from (Miwa, 2020).

An external quantum efficiency was obtained by dividing the number of generated electrons by the number of incident photons, resulting in $3.1 \times 10^{-3}\%$ (Miwa, 2020).

8.2 Thermal Rectification

An electronic rectification device, like a diode, blocks the flow of electrons in one direction while letting them flow in the opposite direction. In a similar way, a thermal rectifier will restrict the flow of phonons in just one direction while leaving the electrical transport unaffected (Schmotz, 2011). A thermal diode can be the building block of other electrical circuit analogs such as thermal memories, gates and transistors. A thermal diode can be used as a passive device to control heat flow in addition to actively driven thermoelectric designs, improving the figure of merit in energy conversion applications (Schmotz, 2011).

Control of the phonon flow while leaving the electron flow unaffected can be done by making the mean free path of the phonons much longer than the electron mean free path, such that the phonon transport is ballistic while the electron transport is diffusive on a particular length scale (Schmotz, 2011).

In a thermal diode, the symmetry of heat flow has to be broken by directing the phonon motion preferably into one direction. Metallic nanostructures can be used to create a geometrical asymmetry that guides heat in a preferential direction. Figure 8.2 shows a schematic diagram of a geometrical asymmetry made from metallic nanostrips that generate an asymmetry in the thermal conductivity of the device, favoring the flow of heat in a certain direction compared to the opposite direction.

Figure 8.3(a) shows a thermal rectifier device with a metallic line that generates a source of heat due to Joule heating. Numerical simulations performed with COMSOL Multiphysics show that there is increased heat flow toward the pyramidal arrangement of metallic lines that has its base closest to the heat source (Fig. 8.3(b)),

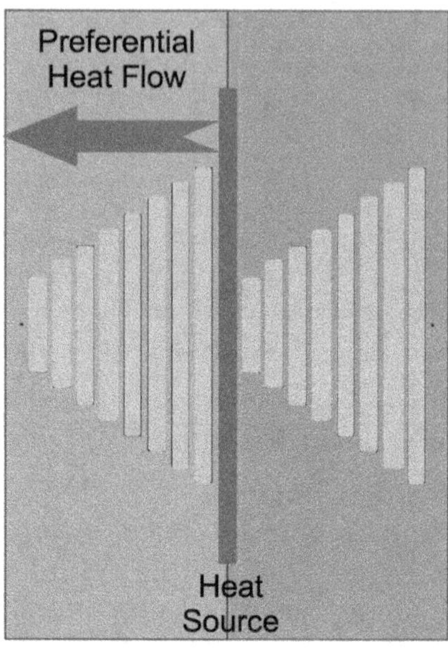

Fig. 8.2. Schematic diagram of a thermal rectifier made from a geometrical asymmetry of metallic nanostrips.

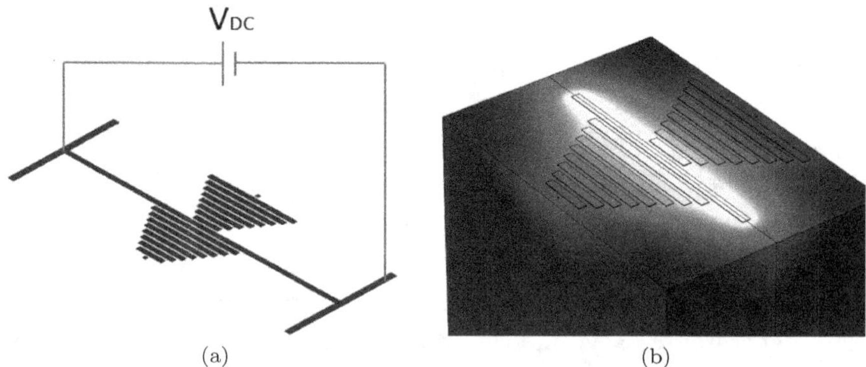

Fig. 8.3. (a) Schematic diagram of a thermal rectifier device with an electric heating line. (b) Numerical simulations of the heat transfer of the thermal rectifier with a heating line.

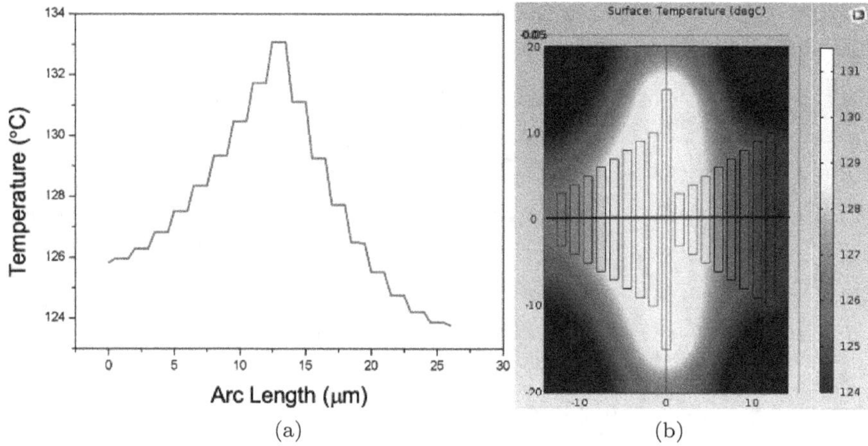

Fig. 8.4. (a) Schematic diagram of a thermal rectifier device with an electric heating line. (b) Numerical simulations of the heat transfer of the thermal rectifier with a heating line.

generating an asymmetrical flow of heat similar to electrons in an electrical diode.

Figure 8.4(a) shows the temperature along a horizontal line through the center of the simulation shown in Figure 8.4(b). The temperature profile shows an asymmetry of about 13 μm of arc length, with a temperature of 126°C at 0 μm arc length and of 124°C at 26 μm arc length.

8.3 Thermoelectric Nanoantennas

Resonant subwavelength nanostructures have been used as an energy source for voltage generation by taking advantage of the heat induced in the nanostructure due to plasmon decay and the thermoelectric effect (Mauser, 2017). By using an antenna as the resonant structure, these nanostructures can take advantage of its polarization sensitivity, directivity, small footprint, tunability and the possibility of integrating it into electronic and photonic circuits [2].

By changing their shape and size, antennas can be tuned to detect optical and infrared frequencies [3–5]. This has attracted attention in the energy harvesting area due to the possibility of designing antennas that harvest electromagnetic wavelengths with lower energy than the energy bandgap of available photovoltaic technology [6]. Among these wavelengths are the infrared and far-infrared portions of the electromagnetic spectrum, where energy can be harvested from the solar spectrum or from thermal sources [7].

Previous work on harvesting thermal or solar energy using antennas was developed by using nanoantennas coupled to high-speed rectifiers, usually known as nano-rectennas [7,8]. These devices operate by converting the AC current induced in the antenna, with frequencies in the order of tens of terahertz, to DC current through rectification. This approach relies on the high switching speeds achieved by tunneling that takes place in metal–insulator–metal diodes [9–11]. Due to the large impedance mismatch between the nanoantenna and the rectifier, the overall efficiency of these devices drop to values down to the $10^{-7} - 10^{-10}$ % range [12].

The possibility of taking advantage of the Seebeck coefficient of the nanoantenna materials to generate voltage would reduce the impedance mismatch [13], which would increase the efficiency of these devices by several orders of magnitude. Several efforts have been made in this direction, for example, in the numerical calculation of the efficiency of these devices [14, 15] and the fabrication and measurement of single-element and arrays of single-metal devices for infrared detection [16].

A bowtie nanoantenna has a broad spectral response, and its shape maximizes the temperature difference between the feed of the antenna and the antenna arms, which increases the thermoelectrically generated voltage [6]. Figure 8.5(a) consists of a 9×9 array of 7.5 μm long classical bowtie antennas with a 60° flare angle (Figure 8.5(b)) connected in series with two bond pads for electrical characterization (Figure 8.5(c)). The series connection is established by using 1.3 μm wide lead lines. The distance between the elements

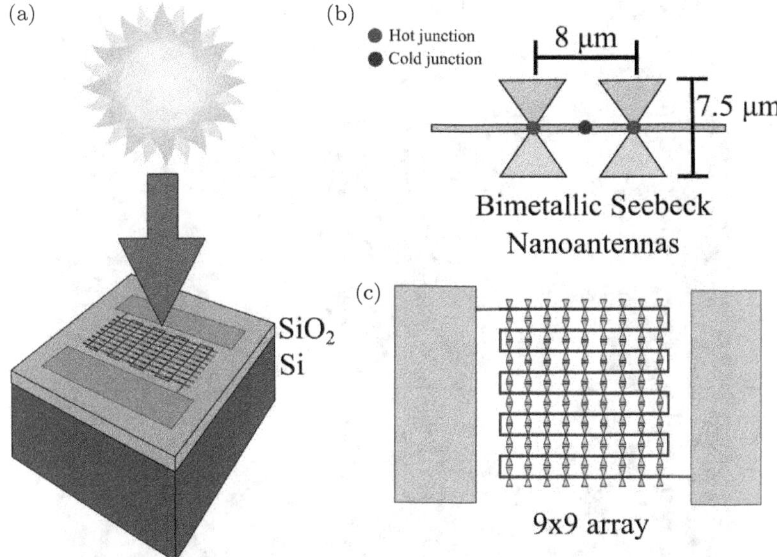

Fig. 8.5. (a) Seebeck Nanoantenna array design for solar energy harvesting. (b) Bowtie nanoantennas used as single elements of the array, and the location of the hot and cold junctions as well as the asymmetry along the current path are shown. (c) 9×9 array of Seebeck nanoantennas.

is 8 μm from center to center. Numerical simulations are performed in order to optimize the distance between elements. The size of the antennas is chosen in order to optimize the response of the device to the far-infrared portion of the electromagnetic spectrum [17], where thermal sources have their highest electromagnetic contribution. However, previous studies have shown that these antennas can also detect visible radiation [10], thus increasing the range of wavelengths that can be harvested using this technology.

The physical mechanism behind the DC current generation in a Seebeck nanoantenna array is thermal diffusion of electrons along the connection line between single-element Seebeck nanoantennas. Figure 8.6 shows the temperature gradient on a pair of bowtie Seebeck nanoantennas. An asymmetry in the connection line along the horizontal axis of each nanoantenna generates a net DC voltage in each nanoantenna and across the whole nanoantenna array.

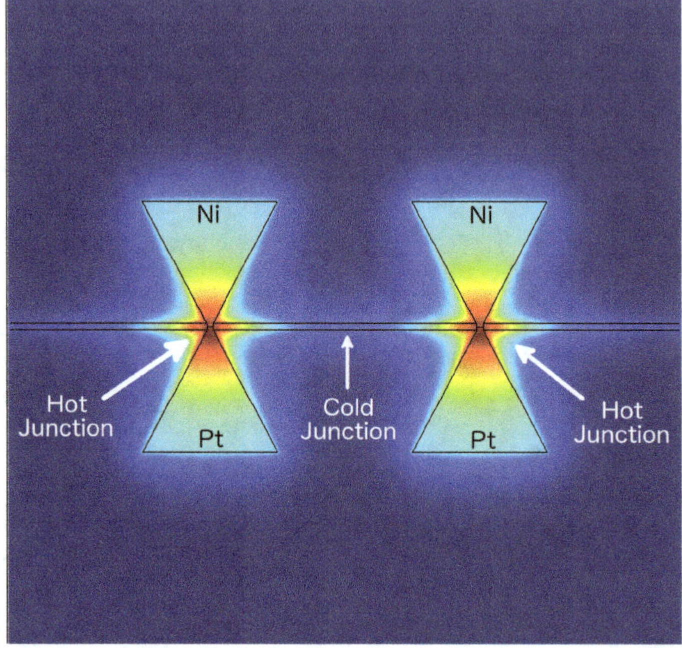

Fig. 8.6. Temperature pattern of a pair of Ni-Pt Seebeck bowtie nanoantennas due to current induced by electromagnetic energy. The figure shows the hot and cold junctions that generate DC current along the connection lines.

Figure 8.5(b) shows the cold and hot junctions in a Seebeck nanoantenna array. The hot junction is generated by Joule heating due to the current induced by electromagnetic energy (Fig. 8.6). By using an antenna, the hot junction is confined to the feed of the antenna, and the cold junction is located between the antennas. This linear configuration of hot and cold junctions reduces the resistance and voltage drop of traditional thermocouples, where the cold and hot junctions are usually located across the thermoelectric materials.

The Seebeck nanoantenna arrays were first evaluated numerically using COMSOL Multiphysics. The simulation procedure consisted of launching a linearly polarized plane wave on the surface of the Seebeck nanoantennas. This plane wave had an irradiance of 1000

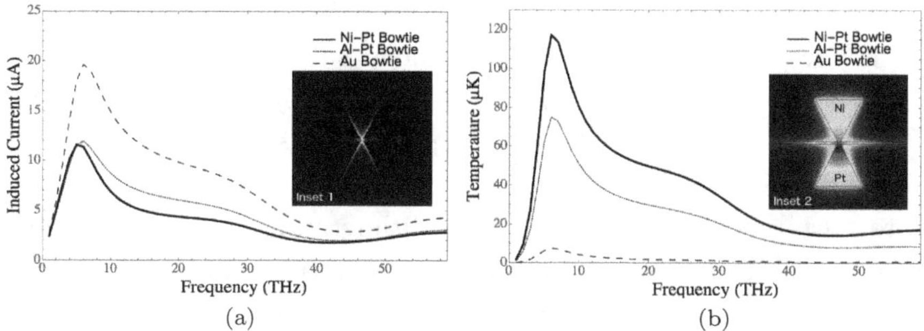

Fig. 8.7. (a) Induced current and (b) temperature at the feed of the nanoantenna due to Joule heating for gold, aluminum-platinum and nickel-platinum Seebeck nanoantennas.

W/m^2. The simulations were performed by doing a frequency sweep of the incident plane wave from 1 to 60 THz. The induced current and the increase in temperature due to Joule heating were calculated by coupling the electromagnetic and the heat transfer behavior through a multiphysics approach.

Figure 8.7(a) shows the magnitude of the current induced in the arms of the bowtie antenna due to the incident electromagnetic field at a 1–60 THz range. From the figure, it can be seen that the peak current induced is at the main bowtie resonant frequency of 6 THz (50 μm in wavelength) in the far-infrared region of the spectrum; however, the response of the antenna extends to frequencies well above 60 THz (5 μm in wavelength).

Inset 1 of Figure 8.7(a) shows the distribution of the induced current on the bowtie nanoantenna, which is symmetric over the two antenna arms. Figure 8.7(b) shows the increase in temperature due to Joule heating produced by the current induced in the nanoantenna arms, and Inset 2 of Fig. 8.7(b) shows the temperature pattern of a Ni–Pt bimetallic bowtie nanoantenna. From Inset 2, it can be seen how the temperature distribution over the nanoantenna arms is asymmetric due to the different conductivity of the materials at the incident frequency. This asymmetry favors the thermoelectric effect, increasing the voltage generated by the thermoelectric nanoantenna.

The fabrication procedure for the three different Seebeck nanoantenna arrays was performed on a silicon substrate with a 300 nm thick SiO_2 layer for electrical and thermal insulation. A double lithographic and metal deposition process was used to create the bimetallic nanoantenna arrays. The nickel, aluminum and gold layers were patterned using electron-beam lithography, and the platinum layers were deposited using a gas injection system and a focused ion beam (FIB) system.

The gold, aluminum and nickel layers were patterned using electron-beam lithography using a conventional lift-off procedure. Briefly, 300 nm of polymethyl methacrylate (PMMA) was spun on the substrates and then baked at 180°C for 15 minutes. Patterning was done with a Raith ELPHY Quantum (Raith America Inc., Ronkonkoma, NY) lithography system, installed in an FEI Inspect F50 field emission scanning electron microscope. The antenna patterns were done using an electron beam exposure of 30 keV and an area dose of 250 $\mu C/cm^2$. After the pattern, exposure development was performed by soaking the samples in an MIBK:IPA solution for 75 seconds. A 50 nm layer of gold, aluminum and nickel was deposited by RF sputtering for each of the different devices fabricated.

Figure 8.8 shows electron micrographs of single elements of the three different types of fabricated Seebeck bowtie nanoantennas.

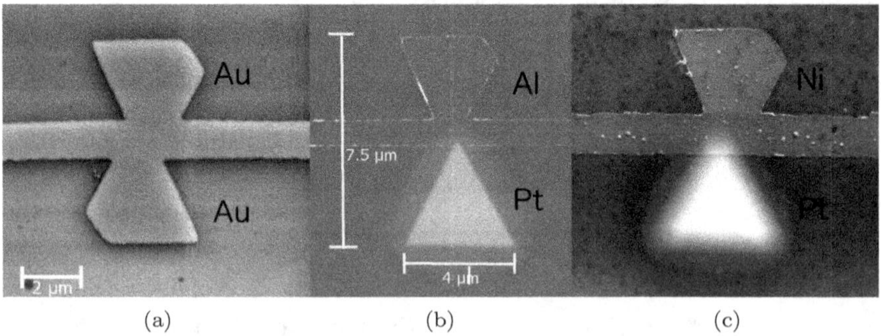

Fig. 8.8. Electron micrograph of single-element Seebeck nanoantennas made of (a) gold, (b) aluminum-platinum and (c) nickel-platinum.

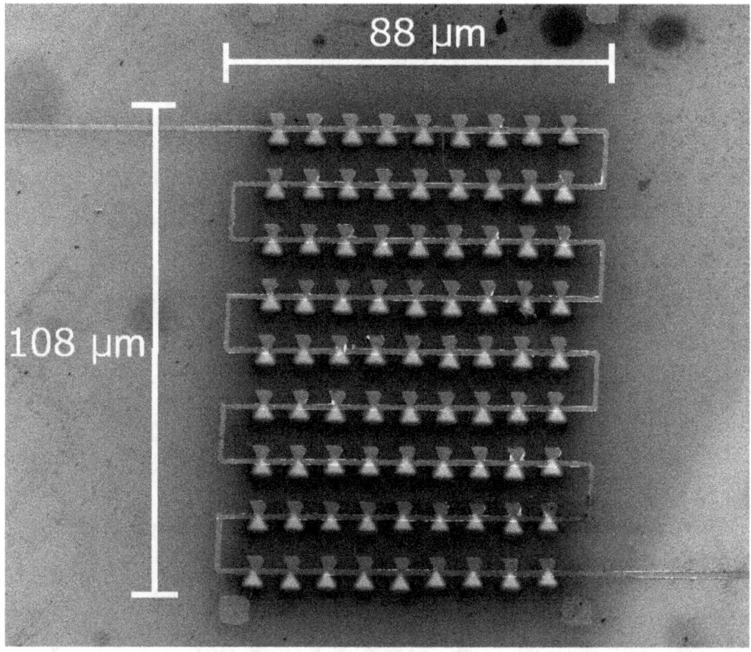

Fig. 8.9.   Electron micrograph of a nickel-platinum Seebeck nanoantenna array.

Figure 8.9 shows a 9 × 9 array of nickel-platinum fabricated Seebeck nanoantennas.

8.4   Plasmonic Heating of Nanostructures

When the size of metallic particles or structures is comparable to or smaller than the wavelength of light, conduction electrons within them will be in phase with the electric field of the incident electromagnetic wave. At certain frequencies, the electromagnetic wave can be at resonance with the oscillations of the conduction electrons. This will lead to an increased absorption of the incoming light, giving rise to a phenomenon called localized surface plasmon resonance (LSPR). This LSPR depends on the size, shape and material composition of the nanostructure. The energy absorbed by these nanoparticles will be released as heat to the environment, making plasmonic nanoparticles efficient light-to-heat converters (Jauffred, 2019).

Heating of plasmonic nanoparticles has become a topic of interest due to their potential for cancer therapy. In this therapy, plasmonic nanoparticles are deposited either by active or passive processes inside tumors and then irradiated by laser light. The increase in temperature of the nanoparticles will kill the tumor cells.

Due to the potential medical applications, most of the research on plasmonic heating of metallic nanostructures has been performed in the near-infrared region of the electromagnetic spectrum: 700–1300 nm. This spectral window is usually referred to as the *therapeutic window* due to the increased transmission of light inside biomedical tissue.

Heating of metallic nanoparticles due to incident light results from Joule heating due to conduction electrons moving inside the nanoparticle. Collisions with lattice atoms result in heat dissipation. The absorption of light by nanoparticles can be characterized by the absorption cross-section, C_{abs}, which is a function of the wavelength and can be directly applied to find the heat power produced by a nanostructure when irradiated at a given wavelength (Jauffred, 2019).

The increase in temperature generated by an irradiated metallic nanoparticle can be calculated with the general heat transfer equation:

$$\rho C_p \frac{\partial T(\vec{r})}{\partial t} = \nabla[\kappa \nabla T(\vec{r})] + Q \qquad (8.1)$$

where ρ and C_p are the density and specific heat capacity at constant pressure, respectively. $T(\vec{r})$ is the absolute temperature, and κ is the thermal conductivity of the surrounding medium. Q is the external source of heat in the medium. This will be the heating generated by the nanostructure and represents the amount of heat produced per unit time and volume in the nanostructure.

The dissipated energy in the nanostructure can be calculated by

$$Q = \iiint q dV, \tag{8.2}$$

where $q = \frac{1}{2} Re(\vec{J}\vec{E}^*)$ is the electromagnetic power loss density within the nanostructure integrated over its whole volume and $\vec{J} = \sigma\vec{E}$ is the current density as a function of the conductivity σ (Jauffred, 2019).

Calculating both the dissipated energy due to the incident electromagnetic field and the increase in temperature can be done by coupling both Maxwell's equations and the heat transfer equation and solving them for the nanoparticle using numerical methods.

An active area of research which involves plasmonic heating of metallic nanostructures is induced cellular hyperthermia for tumor treatments. This involves depositing metallic nanoparticles on cancerous tumors and increasing their temperature using an external electromagnetic field to kill cancerous cells through hyperthermia. To deliver the metallic nanostructures to the tumor, a passive or active approach can be used. In the passive approach, metallic nanostructures are injected into the bloodstream, and they accumulate at the diseased tissue due to its enhanced permeability and retention effect. Alternatively, for an active tumor targeting approach, the nanoparticles can be functionalized with antibodies that specifically target cancer cells. The nanoparticles will bind to the cancerous tissue and can then be activated using an external electromagnetic field, such as a laser. To reach the tumor site, the electromagnetic source must penetrate biological tissue, which can be achieved using a laser with a near-infrared wavelength in the therapeutic window or a longer-wavelength electromagnetic source, such as RF. The metallic nanoparticles must be tuned to resonate at the excitation wavelength.

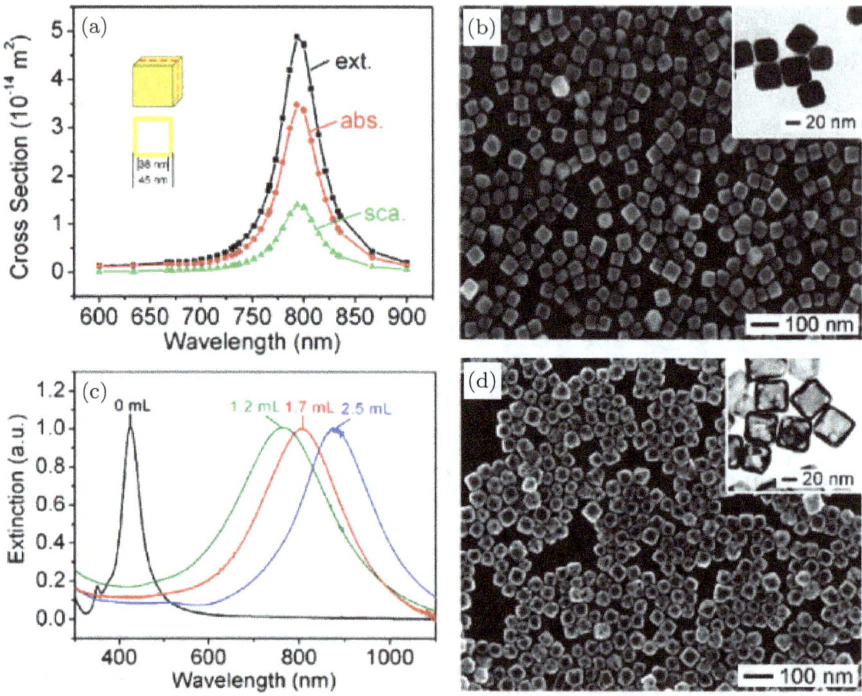

Fig. 8.10.   Numerical simulation (a) and optical characterization (b) of fabricated gold nanocages (b) and (d) as a function of nanoparticle concentration. Taken from (Chen, 2007).

Figure 8.10 shows numerical simulations and optical characterization of gold nanocages for cancerous tumor ablation (Chen, 2007). The nanocages have a hollow interior and a thin gold wall. The plasmon resonance of these structures is at around 800 nm, which falls within the therapeutic window.

Chapter 9

Plasmonic Solar Cells

Solar cells are photovoltaic devices that allow conversion of electromagnetic energy into electrical energy. The use of a renewable source, like the Sun, for the generation of electrical power will reduce the carbon footprint and help combat greenhouse gas emissions and global warming.

Conventional solar cells require trapping the largest amount of solar energy to have high efficiency. In silicon solar cells, light is trapped using a pyramidal surface texture that scatters light into the solar cell over a large angular range, increasing the effective path length in the cell and increasing its efficiency. These pyramidal structures can only be fabricated on thick substrates, increasing the cost of those solar cells.

Thin-film technology could decrease the cost, size and weight of solar cells; however, the absorbance of thin-film devices is small. Plasmonic structures can be used to increase the absorbance of thin-film devices by using any of the following plasmonic effects (Atwater, 2010):

- Metallic nanoparticles can be used as subwavelength scattering elements to couple and trap electromagnetic energy from the Sun.
- Metallic nanoparticles can be used as subwavelength antennas that can couple electromagnetic energy from the Sun into near-field energy.

- A corrugated metallic film on the back surface of a thin photovoltaic absorber layer can couple energy from the sun into SPP modes at the metal/semiconductor interface as well as guide modes in the semincoductor slab.

9.1 Plasmonic Effects in Silicon Solar Cells

Crystalline silicon has an indirect bandgap that limits its light absorption and carrier generation. An indirect bandgap means that the minimum of the conduction band and the maximum of the valence band do not occur at the same momentum value in the crystal lattice. Therefore, an electron transition from the valence band to the conduction band cannot occur by absorbing only a photon; it also needs a phonon to conserve momentum. This reduces the probability of carrier generation, thus reducing the photoelectric efficiency. For a close to 100% absorption efficiency of light, a large silicon film thickness of at least hundreds of microns would be needed (Jong, 2016).

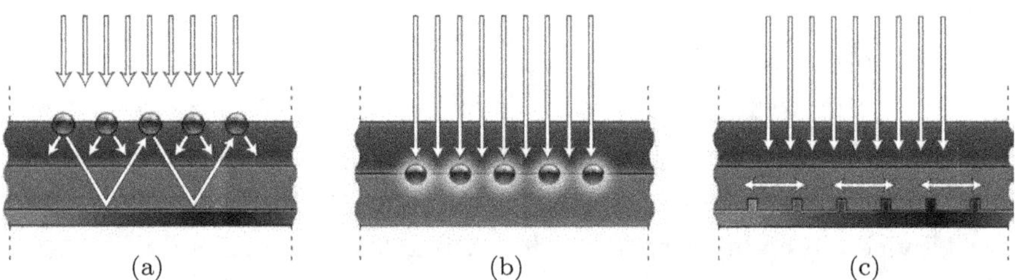

Fig. 9.1. Plasmonic light-trapping geometries for thin-film solar cells: (a) scattering elements that trap electromagnetic energy in the active region of the solar cell; (b) electromagnetic trapping due to localized surface plasmons of metallic nanoantennas embedded in the semiconductor; (c) light trapping by the excitation of surface plasmon polaritons at a metal/semiconductor interface. Taken from (Atwater, 2010).

Much effort has been made to integrate plasmonic structures into silicon solar cells to enhance light absorption. Plasmonic structures can be used to trap more light into the Si film or induce stronger near-field coupling (Jong, 2016).

Plasmonic structures can be placed on the illumination side of the solar cell to take advantage of Mie-type forward scattering, along with an antireflection effect. When the plasmonic structures are placed on the back of the solar cell, backward scattering is tuned to increase the amount of light that goes through the silicon film.

Tuning of light scattering can be done by optimizing the size and shape of metallic nanostructures. An increase in efficiency and photocurrent generation has been achieved by depositing nanoparticles in the top contact of silicon solar cells. Derkacs *et al.* reported an increase of at least 8% in conversion efficiency of a-Si:H solar cells by depositing gold nanoparticles at relatively modest densities (Derkacs, 2006).

Research has demonstrated that depositing spherical gold nanoparticles, ranging from 50 to 100 nm in diameter, onto crystalline silicon p–n junction photodiodes enhances light absorption across a wide spectral range. This improvement is attributed to the interaction between the incoming electromagnetic radiation and the surface plasmon polariton modes in the nanoparticles (Derkacs, 2006).

9.2 Plasmonic-Excitonic Solar Cells

Plasmon enhancement in solar cells has been aimed primarily at improving the efficiency in the visible region of the electromagnetic spectrum. This has been achieved mainly through improved coupling of visible light into the solar cell. The visible-wavelength emphasis comes from the fact that the plasmon resonance of gold and silver is in the visible region of the electromagnetic spectrum.

Since half of the Sun's power reaching the Earth lies in the infrared region of the electromagnetic spectrum, it would be desirable to harvest or improve the collection efficiency of that region that has been usually neglected.

Colloidal quantum dot (CQD) solar cells have the advantage that they can be tuned for different wavelengths. One of the interesting features of CQD technology is that they can generate multiple electron–hole pairs (excitons) from a single photon, which can potentially exceed the Shockley–Queisser limit for traditional solar cells, leading to higher efficiencies.

Currently, CQD technology remains limited by its incomplete absorption of light in the infrared. A near-field enhancement of the photocarrier generation rate of plasmonic nanoparticles could improve the efficiency of CQD solar cells.

Paz-Soldan *et al.* investigated spherical dielectric–metal core–shell nanoparticles, also known as nanoshells, to increase the effective absorption length for NIR photons to length scales much larger than the absorbing film thickness (Paz-Soldan, 2013).

From Figure 9.2, it can be seen that the extinction (absorption + near- and far-field scattering) cross-section is three to five orders of magnitude larger than that of either spherical nanoparticles or nanorods (Figs. 9.2(a) and 9.2(b)). Figure 9.2(d) shows the measured extinction spectrum of nanoshells in methanol solution, consisting of an LSPR centered at 800 nm with a full-width at half-maximum of 280 nm. Due to the presence of a thin metallic shell ($\sim$15 nm), the optical interaction volume of these particles is much larger, reducing the areal density required to scatter incident light completely while minimizing absorption.

Results reported by Paz-Soldan *et al.* show an overall power conversion efficiency (PCE) enhancement of 11% over a nonplasmonic device (PCE = 6.9% vs. 6.2% for the control).

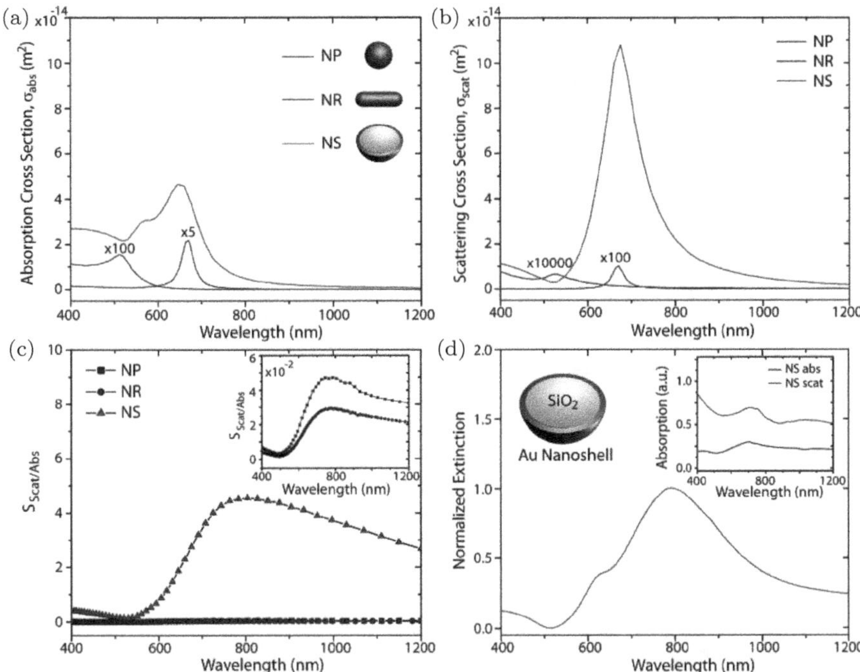

Fig. 9.2. Absorption (a) and scattering (b) cross-section spectra as a function of particle size and shape. The scattering cross-sections take into account both near- and far-field effects. Spherical nanoparticles (NP; diameter = 20 nm) have a limited response beyond $\lambda = 600$ nm, while nanorods (NR; diameter = 10 nm, length = 40 nm) and nanoshells (NS; core diameter = 120 nm, shell thickness = 15 nm) have tunable localized surface plasmon resonances (LSPRs) in the infrared wavelength region. A medium of index 1 was used. The nanoparticle and nanorod spectra are scaled, as noted next to the curves for visual clarity. (c) Calculated scattering-to-absorption ratios (S) showing that nanorods and nanoparticles of commonly synthesized sizes are absorptive, while nanoshells have broadband external field enhancement which exceeds parasitic absorption. The inset shows the same data for nanorods and nanoparticles on a smaller scale. (d) Experimental extinction spectrum of nanoshells in a methanol solution. Insets: schematic diagram of a gold nanoshell cross-section (left) and measured scattering and absorption of nanoshell films (right). Taken from (Paz-Soldan, 2013).

Chapter 10

Metasurfaces

10.1 Introduction

Metamaterials are arrays of special shaped elements that give rise to non-traditional optical, acoustic and heat flow properties that are not attainable with naturally available materials. These metamaterials are composed of periodic subwavelength metallic/dielectric structures that resonantly couple with the electric and magnetic fields of the incident electromagnetic fields. Optical metamaterials can result in negative-index media, zero-index materials and ultrahigh-index materials (Hu, 2021). Materials with such properties can enable new applications in imaging, sensing, data storage, quantum information processing and light harvesting.

A common way to generate artificial structures is by taking advantage of the plasmon resonance of metallic nanostructures. The plasmon resonance will generate a sharp absorption feature in the transmission spectra of the artificial structure. The plasmon resonance is closely related to the size and shape of metallic nanostructures and the dielectric properties of the surrounding medium (Petti, 2016). These metallic nanostructures can be fabricated using nanofabrication tools, such as electron beam lithography, nanoimprint lithography and nanoscale 3D printing.

Metamaterials have paved the way for new applications in the area of nanophotonics. These artificial electromagnetic materials provide innovative ways to control light (Petti, 2016). When metamaterials are designed as two-dimensional (2D) structures, low-cost manufacturing techniques such as nanoimprint lithography can be used for their fabrication. These planar structures can be coupled with on-chip nanophotonic devices for opto-electronics, ultrafast information technologies, microscopy, imaging and sensing. Two-dimensional metamaterials, usually known as metasurfaces, allow the possibility of fully controlling light with planar elements, which has several advantages, including wafer-scale processing and planar photonic integration (Kildishev, 2013).

10.2   2D Metamaterials

Two-dimensional metamaterials, or metasurfaces, are planar, ultra-thin metamaterials which have properties that are not achievable with bulk multilayer materials. Optical metasurfaces can be used to fully control light with planar metamaterial elements and generate "planar photonics." In practice, these metasurfaces are achieved using a patterned metal–dielectric layer that is very thin compared to the wavelength of the incident light and is typically deposited through lithography on a supporting substrate.

Effective optical properties have been found to deviate from classical reflection and refraction laws; therefore, the response of metasurfaces is different from the experimental response of bulk materials. To design flat photonic devices, full-wave electromagnetic simulations are usually utilized; however. semi-analytical models such as the transmission line model (TML) can be used to determine the effective permittivity and effective permeability of the metasurface. Also, machine learning (ML) techniques have been used to design and optimize the performance of metasurfaces (Ghorbani, 2021).

Plasmonic metasurfaces consisting of arrays of metallic nanostructures are strong candidates for sensing, spectroscopy and lasing applications.

Metasurfaces fabricated from antenna arrays have been used in communication applications (Pozar, 1997) (Sievenpiper, 1999) (Ryan, 2010) or as highly confined cavity resonators (Caiazzo, 2004) (Holloway, 2008).

An antenna array can be configured as a metasurface. The antenna elements in reflectarrays and transmitarrays act as phase-controlling resonators for manipulating the direction in which electromagnetic waves are received or emitted. A phase discontinuity along an interface can be engineered using an antenna array to fully steer light or to accomplish anomalous reflection and refraction (Fig. 10.1).

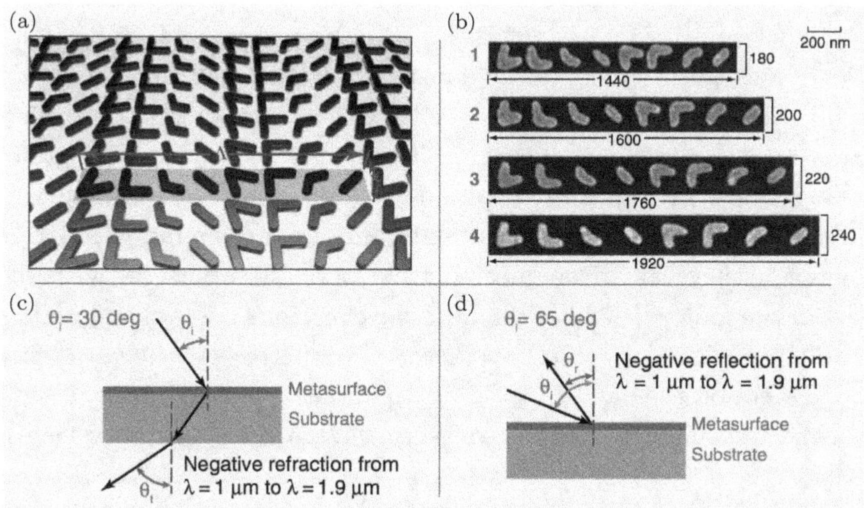

Fig. 10.1. (a) Schematic diagram of a V-shaped nanoantenna array metasurface. (b) Fabricated V-shaped nanoantenna arrays with different periodicity. (c and d) Negative refraction and reflection angles obtained experimentally from the V-shaped nanoantenna arrays. Taken from (Kildishev, 2013).

Metasurfaces can be engineered to control the phase of a light wave similar to lenses and mirrors but using an extremely thin device. Applications of metasurfaces are very diverse: they can be used for spatial phase modulation, beam shaping, beam steering and plasmonic lenses, to name a few.

10.3 Perfect Lenses

Optical systems are fundamentally limited by diffraction, and the ability to surpass this fundamental limit has attracted considerable attention due to the great benefits that could bring to diverse fields such as bio-imaging, data storage and lithography. Subwavelength resolution has been demonstrated with scanning near-field optical microscopy (NSOM); however, this technique does not project a whole image like a regular lens does.

Left-handed materials (LHMs) have been proposed by Sir John Pendry as means to obtain super-resolutions below the diffraction limit. LHMs are materials in which the permittivity and permeability are both simultaneously negative. The phase velocity and the group velocity are in opposite directions, so the index of refraction is negative. Such negative-index materials are not found in nature; however, they can be engineered from periodic structures with the size of the unit cell smaller than the wavelength. These metamaterials are characterized by effective material properties which are highly dependent on the incident wavelength and can be engineered to have a negative effective index of refraction for a particular wavelength band.

In 2000, Sir John Pendry reported in *Physical Review Letters* that a slab of negative-refractive-index material could focus all the Fourier components of a 2D image, even those that do not propagate in a radiative manner, and that a version of this lens operating at the frequency of visible light could resolve objects as small as a few nanometers (Pendry, 2000).

10.4 Metasurface Cloaking

The use of metamaterials makes it possible to mold the flow of light. These metamaterials are designed from resonant periodic structures considerably smaller than the wavelength. The flow of light can be redirected around objects so that these objects would seem invisible.

The first practical realization of an invisibility cloak using metamaterials was reported in the journal *Science* in 2006 (Schurig, 2006). In this work, an artificially structured metamaterial designed to have electric and magnetic resonances over a narrow frequency band was fabricated for the microwave region of the electromagnetic spectrum. Further, a copper cylinder was made effectively invisible to incident microwave radiation. Figure 10.2(a) shows a finite element simulation of the electric field pattern of an incident electromagnetic wave hitting a cloaking device. Figure 10.2(b) shows the fabricated cloaking device in the microwave region along with the elements used to fabricate the metasurface.

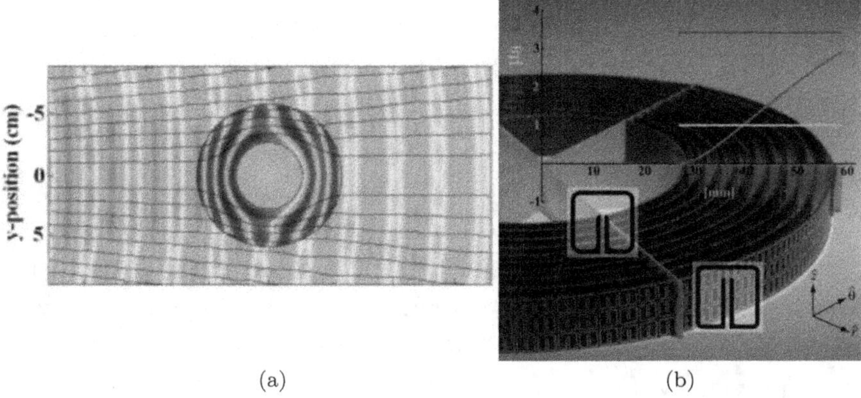

(a) (b)

Fig. 10.2. (a) Numerical simulation of the electric field pattern of an electromagnetic wave hitting a cloaking device, with the black stream indicating the direction of power flow (i.e., the Poynting vector). The cloak lies in the annular region between the black circles and surrounds a conducting Cu cylinder at the inner radius. (b) Fabricated 2D microwave cloaking structure. Taken from (Schurig, 2006).

The process for designing an invisibility cloak makes use of transformation optics, which consists of performing a coordinate transformation over the space surrounding the volume that needs to be concealed. Since Maxwell's equations are form-invariant to coordinate transformations, only the components of the permittivity tensor ϵ and the permeability tensor μ are affected by the transformation. By implementing these material properties, the concealed volume plus the cloak appear to have the properties of free space when viewed externally (Schurig, 2006).

10.5   Design of Metasurfaces based on Artificial Intelligence

Traditional design approaches for the design of metamaterials consist of numerical modeling, trial-and-error, parameter sweep and optimization algorithms (Ghorbani, 2021). Full-wave numerical simulations assisted by optimization algorithms consume a large amount of computer resources and are time-consuming. Also, if the design requirements change, the whole process must be repeated. Machine learning can be used in the design of metasurfaces by automatically learning the connection between the design requirements and the metasurface design by inferring the connection from examples of past designs.

Machine learning can be used to design metasurfaces by training a computational platform to deduce fundamental principles based on previously designed metasurfaces. Deep learning is a particular branch of machine learning in which an artificial neural network (ANN) with a large number of layers is used to generate deductions from previously fed data. An ANN is an interconnected group of nodes inspired by the brain structure and the connection of neurons. It consists of four main components: inputs, weights, a bias or threshold and an output. The nodes, also called neurons, of an ANN are organized into multiple layers, where neurons of one layer connect only to neurons of the immediately preceding and immediately following layers. The layer that receives external data is the *input layer*, while the layer that produces the output is the *output layer*, and in between them, there are zero or more hidden layers.

Chapter 11

Metallic Nanostructures for Field Enhancement Spectroscopy

The plasmonic effect, resulting from the interaction of light with metallic nanoparticles, leads to an enhancement of the electric field. This enhancement can be utilized to improve processes that depend on the electric field, including the Raman effect, fluorescence and infrared absorption. To this end, specialized substrates have been developed, consisting of metallic nanostructures, with the aim of enhancing these processes and increasing the sensitivity of associated characterization techniques, such as Raman, infrared and fluorescence spectroscopy. These techniques, when improved by metallic nanoparticles, are differentiated from their non-enhanced counterparts and are commonly referred to as surface-enhanced Raman spectroscopy (SERS), plasmon-enhanced fluorescence, and surface-enhanced infrared spectroscopy (SEIRA).

11.1 Surface-Enhanced Raman Spectroscopy (SERS)

The Raman effect was discovered by Chandrasekhara Venkata Raman in 1928, who was awarded a Nobel Prize for this discovery in 1930. This effect is due to inelastic scattering of photons by matter. When photons are scattered, the great majority of them are scattered elastically, which means the photons will have the same energy (wavelength) after colliding with atoms.

A very small amount of photons (1 in 1 million) will gain or lose energy after colliding with atoms (inelastic scattering), which means that the scattered light will have a different color than the incident light — this is the Raman effect (Ramirez-Elias, 2018). If the scattered photons lose energy, they suffer a Stokes shift, while if the scattered photons have higher energy than the incident photons, then the energy difference is called an anti-Stokes shift.

These changes in photon wavelength are used to generate a spectrum of scattered light intensity as a function of frequency. This Raman spectrum has the vibrational mode information of the molecules that scattered the incident light. Each molecule has a distinct Raman spectrum, which gives a molecular fingerprint that can be used to identify the composition of the scattering material. Due to the small number of photons that experience inelastic scattering, the Raman signal is usually very small and not effective in identifying molecules in low concentrations.

In the 1970s, surface-enhanced Raman spectroscopy (SERS) was developed. In this technique, the Raman spectrum is significantly amplified (with enhancements of 10^6) thanks to metallic nanoparticles. Using this technique, it is possible to detect molecules in low concentrations through Raman scattering.

The main mechanism that generates SERS is the field enhancement due to plasmon resonances in metallic nanostructures. Additionally, there exists polarizability enhancement due to chemical effects. The plasmon enhancement is well known, and it is given for each molecule by (Schatz, 2006)

$$enh = |\vec{E}(\omega)|^2 |\vec{E}(\omega')|^2, \tag{11.1a}$$

where $\vec{E}(\omega)$ is the local field enhancement at the incident wavelength and $\vec{E}(\omega')$ is the field enhancement at the Stokes-shifted frequency ω'. The enhancement enh is often approximated by assuming $\vec{E}(\omega) \approx \vec{E}(\omega')$; therefore, $enh \approx |\vec{E}(\omega)|^4$. This approximation is valid since, usually, the plasmon width is often large compared to the Stokes shift.

Localized surface plasmons (LSPs) concentrate the light's energy at hot spots on the surface of metallic nanostructures; therefore, they are ideal for enhancing Raman scattering processes. Enhancement can also be obtained using a sharp metal; this technique is called tip-enhanced Raman spectroscopy (TERS). As a result of the enhancement, due to the LSPs, the detection level of Raman scattering is getting closer to the single-molecule detection level.

Plasmonic nanostructures used in SERS have opmimal shapes and sizes that enhance Raman scattering. Enhancement factors at the level of 10^6 are common with spherical gold nanoparticles. If the nanoparticles have sharp corners, this can give a higher enhanced field. Using nanoneedles or urchin-type particles, giant Raman enhancements as strong as 10^{10} can be achieved. Also, metallic nanogap elements, such as nanoantennas, have been used, which give high Raman enhancement factors.

11.2 Surface-Enhanced Infrared Absorption (SEIRA) Spectroscopy

Similar to Raman spectroscopy, infrared absorption spectroscopy provides chemical information about a sample by identifying vibration bands that are characteristic of every molecule. These lattice molecular vibrations constitute a spectroscopic fingerprint of a molecule and can be used for its detection.

Also, similar to Raman spectroscopy, the detection of low amounts of molecules is challenging due to low signal stress. By using a surface enhancement effect, the signal intensity can be increased by several orders of magnitude. The enhancement mechanism is due to the plasmonic field of the metallic surface; this plasmonic field decays sharply in intensity with distance from the surface.

The topology of the metal surface significantly contributes to the enhancement properties. Typically, roughened surfaces of nanometer scale are used. Since the average distance between metal particles at a given roughness is much smaller than the wavelength of IR radiation, the optical properties of the metal film are assumed to arise from a mixed composition of metal, adsorbed molecules and the surrounding host medium.

Most of the SEIRA substrate structures that have been investigated are focused on rod- or rod-dimer-based antennas, whose aspect ratio is tailored so that their resonance matches the specific vibrational frequency of interest of the analyte molecules. Other designs include split-ring Fan structures, asymmetric metamaterials and log-periodic antennas. In general, these examples exploit capacitive coupling between adjacent structures to maximize field confinement and enhancement at the hot spot. The hot spot can exhibit even higher near-field enhancement by coupling plasmonic structures in narrower gaps. Nanosized gaps have been fabricated using electron-beam lithography (EBL), focused ion-beam (FIB), electromigration, nanospherelithography, photochemical metal deposition and template stripping. Nevertheless, reproducible fabrication of an arbitrary plasmonic structure with nanoscale gaps remains a substantial challenge at present.

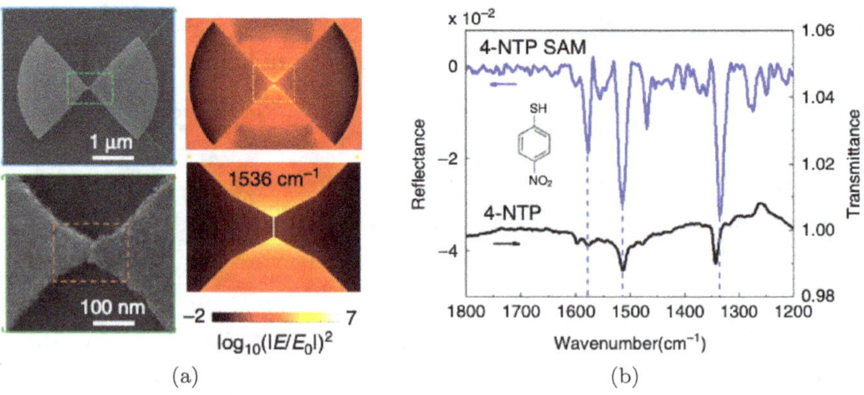

Fig. 11.1. (a) SEM and simulated electric field distribution of a bowtie antenna (at $1536cm^{-1}$) (b) SEIRA of 4-NTP SAM on a single antenna. Reference spectra of solid-state 4-NTP are shown in black with prominent vibrations indicated by dashed lines. From (Dong, 2017) and (Wang, 2021).

In (Dong, 2017), the SEIRA responsivity of a bowtie plasmonic antenna is investigated. In this work, the design incorporates a sub-3 nm gap that is positioned above a gold film with a silica spacer layer. The extremely large field enhancement achieved with this antenna design allowed the detection of very small numbers ($\sim$500) of aromatic-derivatized thiol molecules on the antenna using a standard commercial FTIR spectrometer.

11.3 Metal-Enhanced Fluorescence

Molecules in an excited state can emit photons. After electronic excitation by ultraviolet or visible radiation, atoms and molecules can undergo thermal or radiative deactivation processes before relaxing to the ground state. They can emit photons with longer wavelengths than the incoming exciting radiation, that is, they can fluoresce in the UV-vis-near-infrared (NIR) range. The study of fluorescence relaxation processes is one of the experimental bases on which modern theories of atomic and molecular structure are founded. Technological improvements in both optics and electronics have greatly expanded fluorimetric applications, particularly in analytical fields, because of the high sensitivity and specificity afforded by the methods.

Metal-enhanced fluorescence (MEF) has been applied in many chemical and biological systems where the fluorescence signals are amplified with increased photostability. Although MEF has been extensively investigated, the details of the MEF mechanism are not clearly understood yet. The increased local electric field near ($<$60 nm) the plasmonic nanoparticles amplifies both the absorption and emission of the fluorophore, which is often called the electric field effect of MEF. The decrease in the radiative lifetime of the fluorophore, which was first proposed as the induced plasmon effect of MEF, has recently been understood as the surface plasmon coupled emission (SPCE) or the fluorophore radiation through the scattering mode of the nanoparticles.

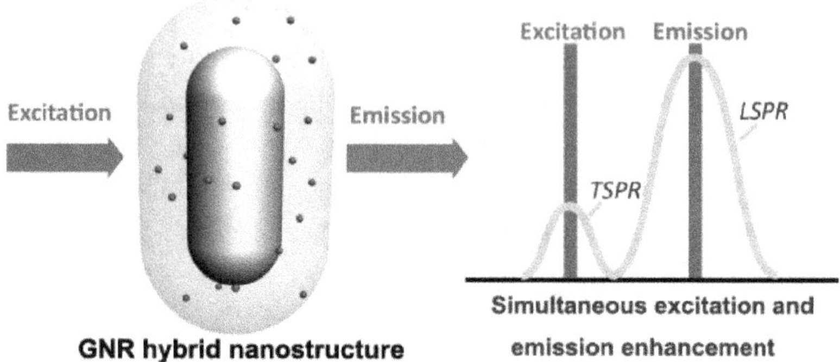

Fig. 11.2. Simultaneous emission and excitation enhancement due to transversal and longitudinal surface plasmon modes of gold nanorods. From (Liu, 2013).

Fluorescence from a fluorophore can be characterized in terms of quantum yield and lifetime. As with other field enhancement techniques, the LSP can be tuned by changing the size, shape and material of the metallic nanoparticles to match either the peak excitation or the peak emission of the fluorophore. For a plasmon resonance close to the peak excitation wavelength, the light will be confined and a high excitation efficiency can be expected. If the plasmon resonance is close to the peak emission wavelength, then the highest emission efficiency will take place for the molecules close to the metallic nanoparticle.

Research has been done to fabricate metallic nanoparticles with double plasmon resonances. In metallic nanorods, for example, there is a longitudinal and transverse plasmonic resonance corresponding to the length and width of the nanorods. These resonances can be tuned to match the emission of excitation wavelengths of the fluorophores, increasing the fluorescence efficiency (Fig. 11.2).

Chapter 12

Optical Antennas

12.1 Introduction

Antennas are fundamental devices for RF, microwave and millimeter wave wireless communication systems. The term *antenna* was adopted from the Latin language, where it was used to describe the post that supports the sail in a sailing vessel. This term is now used to describe the sensing elements located in the head of some insects, and it was used by Gugliermo Marconi to describe the vertical metallic rods he was using to transmit and receive electromagnetic waves in his 1901 transatlantic experiment from Poldhu, UK, to St Johns, Newfoundland (Boswell, 1997).

A common definition of an electromagnetic antenna is a device that converts a time-varying electrical current into a time-varying electromagnetic wave. This device can also work in its reverse operation mode, where an electromagnetic wave can be converted into an electrical current. Antennas can work in emission mode, where an electromagnetic wave is generated, or in reception mode, where an electrical current is induced due to an electromagnetic wave that impinges on the antenna.

An antenna is usually formed by resonant metallic structures. When an electromagnetic wave falls on the antenna, a current is induced in the metallic structures. This induced current can be detected by a sensor located at the feed of the antenna. If the antenna is used as an emitter, then a current generator located

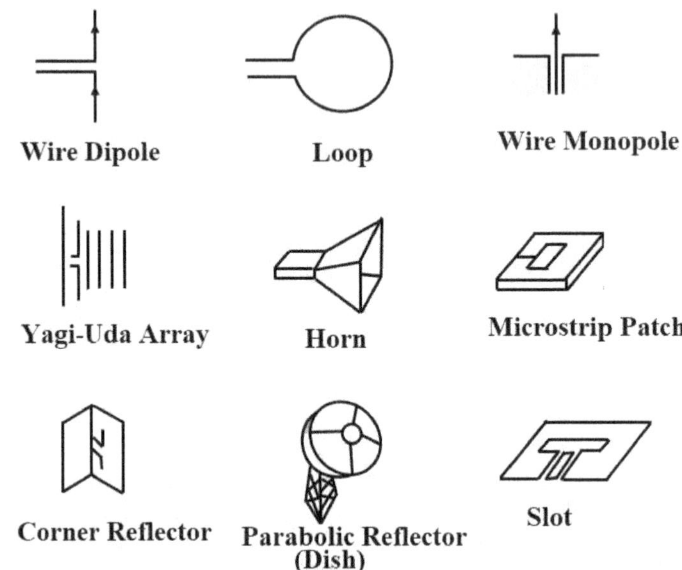

Fig. 12.1. Different types of antennas. Each has characteristic frequency bandwidth and polarization.

at the feed of the antenna will radiate an electromagnetic wave with the same frequency as the generator. The antenna elements are designed to resonate at a particular frequency or wavelength, so that the incident electromagnetic waves at that particular frequency would be preferentially detected or emitted. Depending on the antenna structure, different polarization, bandwidth and radiation characteristics can be obtained. Figure 12.1 shows several types of antennas.

The size of the antenna elements is proportional to the desired resonant wavelength. So, at RF and microwave frequencies, where the frequencies range from 3 kHz to 300 MHz and 300 Mhz to 300 GHz, respectively, the size of antennas at those frequencies would be in the millimeter to meter range. Since Maxwell's equations and electromagnetic wave theory work for any wavelength, reducing the size of the antenna elements would make electromagnetic detection and emission possible at shorter wavelengths, as well as even optical wavelengths, which are in the nanometer range (400–700 nm).

Advances in nanofabrication had made it possible now to fabricate arbitrary metallic structures with nanometer resolution, which allows the practical implementation of nanoantennas for optical wavelengths.

12.2   Antenna Theory

Microstrip and printed circuit antennas have gained prominence as viable antenna elements and arrays. These types of antennas have several advantages such as low profile, low cost, light weight, conformity to surface, mass production, variable-frequency operation possibilities and compatibility with integrated circuit technology. Their limitations are low gain and narrow bandwidth. The main difference between microstrip and printed circuit antennas is that the first ones have a ground plane which affects the radiation characteristics of the antenna significantly. This ground plane can be modeled using image theory, that is, introducing virtual sources that will account for the reflections occurring at the ground plane.

A free-space antenna radiates equally on both sides because of its symmetry. The presence of a dielectric substrate will break the symmetry and change the current distribution and wave velocity of the antenna. One of the main features of antennas on dielectrics is that they tend to radiate more power into the dielectric in the ratio $\varepsilon_s^{3/2}$: 1, where ε_s is the relative electrical permittivity of the substrate. Also, waves propagate differently along metals at a dielectric interface. The waves tend to propagate at a velocity that is intermediate between the velocity of waves in the air and the velocity of waves in the dielectric. For antennas on a substrate in the millimeter-wave range, waves propagate with a velocity close to that characteristic of a material with a dielectric constant equal to the mean of the two dielectric constants. This is a quasi-static approximation to the effective permittivity (Rutledge, 1983) and is only valid for frequencies below a few GHz; above these frequencies, the frequency dispersion of the effective permittivity must be taken into account. Hasnain *et al.* (Hasnain, 1986) derived an analytical expression for the dispersion of the effective permittivity, which is given by

$$\sqrt{\varepsilon_{\textit{eff}}} = \sqrt{\varepsilon_q} + \left(\frac{\sqrt{\varepsilon_s} - \sqrt{\varepsilon_q}}{1 + a(\frac{f}{f_{TE}})^{-b}} \right), \qquad (12.1)$$

where ε_q is the quasi-static value of the permittivity given by $\varepsilon_q = \frac{\varepsilon_s + 1}{2}$, f is the frequency of the wave in Hz and f_{TE} is the cut-off frequency for the lowest-order TE mode, which depends on the substrate thickness d and is given by

$$f_{TE} = \frac{c}{4d\sqrt{\varepsilon_s - 1}}, \qquad (12.2)$$

where c is the velocity of light in a vacuum. The parameter a depends on the configuration and dimensions of the transmission line. For the specific case illustrated in (Hasnain, 1986) $a \approx 51$, b was found to be always around 1.8 independent of the dimensions. Figure 12.2 shows the effective permittivity as a function of substrate thickness for waves at 10.6 μm wavelength (28.3 THz). We can see how, for a substrate thickness below 1 μm, the effective permittivity is equal to the quasi-static value, and for a thickness larger than 100 μm, the effective permittivity is equal to the substrate thickness.

The most basic properties of an antenna are its radiation pattern, gain, impedance and polarization. These properties are identical for linear passive antennas used either as a transmitter or a receiver due to the reciprocity theorem. A radiation pattern is a graphical representation of the far-field properties of an antenna. It can be measured by rotating the antenna and plotting the response as a function of angular coordinates. Power gain is defined as 4π times the ratio of the radiation or detection intensity to the net power accepted by the antenna. The input impedance of an antenna is the impedance presented by the antenna at its terminals and is composed

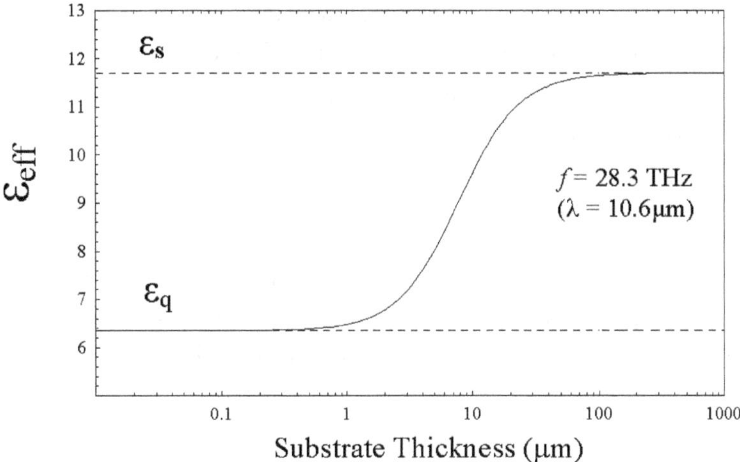

Fig. 12.2. Effective permittivity as a function of substrate thickness of a coplanar waveguide at 28.3 THz (10.6μm).

of real and imaginary parts:

$$Z_{in} = R_{in} + jX_{in}. \tag{12.3}$$

The input resistance, R_{in}, represents dissipation, while the input reactance, X_{in}, represents the power stored in the near field of the antenna. To maximize the power transfer from a transmitter to an antenna, the antenna impedance should be a conjugate match to the impedance of the load, meaning that input resistances should be the same and the reactances should have equal magnitude but opposite signs. Antennas that are electrically small (much smaller than a wavelength) have a large input reactance, in addition to a small radiation resistance. The polarization of a transmitting antenna is the polarization of the wave radiated by the antenna. In the case of receiving antennas, the polarization of the antenna is the polarization of the wave that will maximize its response.

With each antenna, we can associate a number of equivalent areas. These are used to describe the power characteristics of the antenna when a wave impinges on it. One of these equivalent areas is the effective collection area (aperture) A_{eff}, which is defined as the ratio of the available power at the terminals of a receiving antenna (P_{in}) to the power flux density (irradiance) of a plane wave incident on the antenna (W_{in}):

$$A_{eff} = \frac{P_{in}}{W_{in}}, \tag{12.4}$$

with the wave being polarization matched to the antenna (Balanis, 1997). The effective collection area is related to the maximum directivity of an antenna by the following equation:

$$A_{eff} = \frac{\lambda^2 D_{\max}}{4\pi}, \tag{12.5}$$

where the maximum directivity of the antenna is defined as the ratio of the maximum radiation intensity to the radiation intensity averaged over all directions.

12.3   Infrared Antennas

IR detectors can be classified into photon detectors and thermal detectors (Rogalski, 2020). Photon detectors absorb radiation through interactions with electrons, either bound to lattice atoms or to impurity atoms, or with free electrons. Photon detectors show a selective wavelength dependence on the response per unit incident radiation power. They exhibit both good signal-to-noise performance and a very fast response. To achieve these, the photon detectors require cryogenic cooling to prevent the thermal generation of charge carriers. Cooling requirements are the main obstacle to the more widespread use of IR systems based on semiconductor photodetectors, which make them bulky, heavy, expensive and inconvenient to use.

On the other hand, with thermal detectors, the incident radiation is absorbed to change the temperature of the material, and the resultant change in some physical property is used to generate an electrical output. The signal does not depend on the photon nature of the incident radiation. Thus, thermal effects are generally wavelength independent, while the signal depends on the radiant power (or its range of change) but not on its spectral content. Thermal detectors are typically operated at room temperature. They are usually characterized by modest sensitivity and slow response (because heating and cooling of a detector element is a relatively slow process), but they are cheap and easy to use. They have found widespread use in low-cost applications, which do not require high performance or speed.

The work by Matarrese and Evenson (Matarrese, 1970) showed that infrared antennas can be used to collect infrared energy and apply it in the form of a voltage at the infrared frequency to metal–insulator–metal(Fumeaux, 1998) or metal–semiconductor junctions (Tsang, 1977). The effect of the infrared frequency voltage on the junction is to produce a change in the DC voltage or current. This change is due to the rectification of the current at the infrared frequency by the nonlinearity of the junction (Faris, 1975).

The bolometer is a popular thermal detector. It consists of a resistive element constructed from a material with a very small thermal capacity and a large temperature coefficient so that the absorbed radiation produces a large change in resistance. The device is operated by passing a bias current through the detector and monitoring the output voltage. In the case of bolometers, radiant power produces heat within the material, which in turn produces a change in resistance. There is no direct photon–electron interaction.

The name "microbolometer" is usually used when the temperature-sensitive element is much smaller in size than the wavelength (Hwang, 1979). Microbolometers have a smaller thermal mass and are faster than traditional bolometers; however, with traditional bolometers, the resistive element is not only used to detect the radiation, but its surface is also used to collect it.

By coupling an antenna to a small bolometer, it is possible to have fast detectors without sacrificing collection area. The antenna will be in charge of collecting radiation, and the bolometer's only function will be to detect it. Also, when the detector element is small, a small amount of energy is needed to cause a large change in resistance; therefore, a smaller detector will have better responsivity.

These types of detectors are also known as antenna-coupled microbolometers. They work by collecting infrared radiation by the antenna as inducing current in its elements. The generated current has the same frequency as the incident radiation. This generated current will flow through the sensing element — in this case, the microbolometer — and will increase its temperature due to Joule heating. This change in temperature will make the bolometer change its resistance and therefore providing the detection mechanism. Additional advantages of using an antenna as the collection element are its directivity and the possibility of polarization and wavelength selection. Many different types of planar antennas have been coupled to infrared detectors: dipole (Wilke, 1994), bow-tie (Fumeaux, 1998), spiral (Grossman, 1991), log-periodic (Chong, 2003), slot-antenna (Yasuoka, 1998) and microstrip patches (Codreanu, 1999).

12.4   Discrete and Self-Similar Antennas

The use of optical antennas has attracted attention due to the electric field enhancement in subwavelength regions, which has potential applications in nanolithography, biosensing and optical imaging (Yang, 2014).

Optical bowtie nanoantennas are nanostructures that operate at higher bandwidths than traditional dipoles. These structures generate a stronger field enhancement at the gap of the antenna due to surface plasmon currents generated in the antenna arms and strong capacitive coupling at the gap of the antenna (Yang, 2014).

Fractal structures have been used as antennas in order to increase the frequency operation band (Werner, 2003). Fractals are constructed by recursively using an operator on a basis shape. A fractal bowtie antenna is constructed by using a triangle as the basis shape. These fractal structures are also known as Sierpinski gaskets (Cakmakyapan, 2014). These structures show multiple hot spots where the electric field is enhanced and localized, which can be used as efficient SERS substrates (Asiala, 2011) (Hsu, 2010).

The discrete bowtie antenna is based on the traditional bowtie antenna design but made of discrete elements in order to increase the field enhancement near the antenna and increase its frequency response (Moreno, 2020).

Classical and discrete bowtie antennas were patterned using electron beam lithography on silicon substrates (Figure 12.4), and their electric field enhancement was investigated using near-field scanning optical microscopy and finite element simulations. The polarization dependence, frequency dependence and electric field enhancement spots were compared through numerical simulations with those of fractal antennas. The fractal antennas used for comparison were constructed by using triangles as basis shapes and subtracting inverted triangles from the original basis shape. Numerical simulations were performed on two fractal structures obtained by iterating one and two times the subtraction procedure.

Finite element simulations were performed using COMSOL Multiphysics by launching a linearly polarized plane wave at a 60° angle of incidence and evaluating the electric-field (E-field) at the antenna plane. Results were obtained by varying the polarization and frequency of the incident electromagnetic wave from 1 to 100 THz (3–300 μm in wavelength).

The fabrication process for the classic and discrete nanoantennas was performed using a standard electron-beam lithography and lift-off procedure. The devices were fabricated on Si wafers with 300 nm of SiO_2 as the thermal and electrical isolator. The fabrication procedure used was the following: 300 nm of polymethyl methacrylate (PMMA) was spun on the substrates and then baked at 180° for 15 minutes. Then, patterning was done with a Raith ELPHY Quantum (Raith America Inc., Ronkonkoma, NY) lithography system, installed in an FEI Inspect F50 field emission scanning electron microscope. The antenna patterns were done using an electron beam exposure of 30 keV and an area dose of 250 $\mu C/cm^2$. Development was performed by soaking the samples in an MIBK:IPA solution

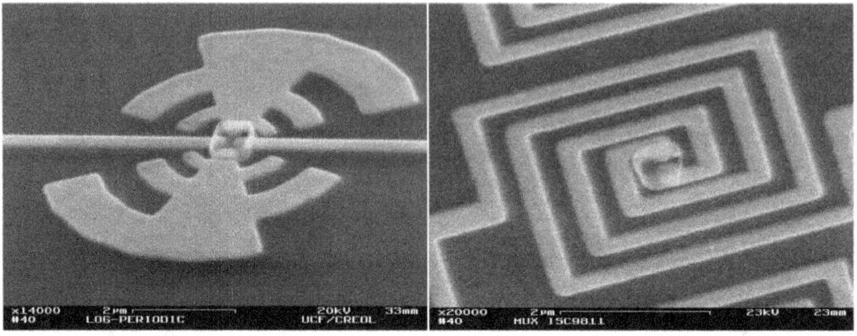

Fig. 12.3. Log-periodic and spiral antenna coupled to a Nb microbolometer.

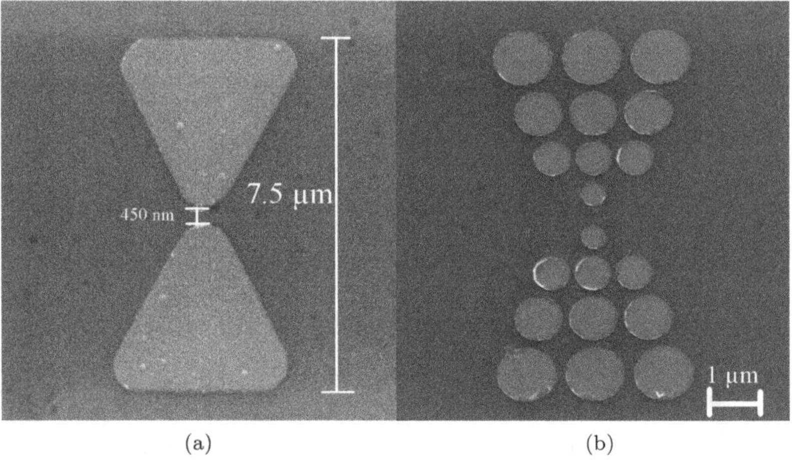

(a)   (b)

Fig. 12.4. (a) classical bowtie antenna and (b) discrete bowtie antenna fabricated on silicon substrates.

for 75 seconds, and 50 nm of gold were deposited by RF-sputtering and left on an acetone bath overnight. Figure 12.3 shows scanning electron micrographs of the fabricated structures.

Near-field measurements were taken using a custom-built scanning-scattering near-field optical microscope (s-SNOM) operating in the pseudo-heterodyne mode of detection (Ocelic, 2006).

Figure 12.5 shows the electric field at the gap of the classic and discrete bowtie antennas as a function of frequency. From the

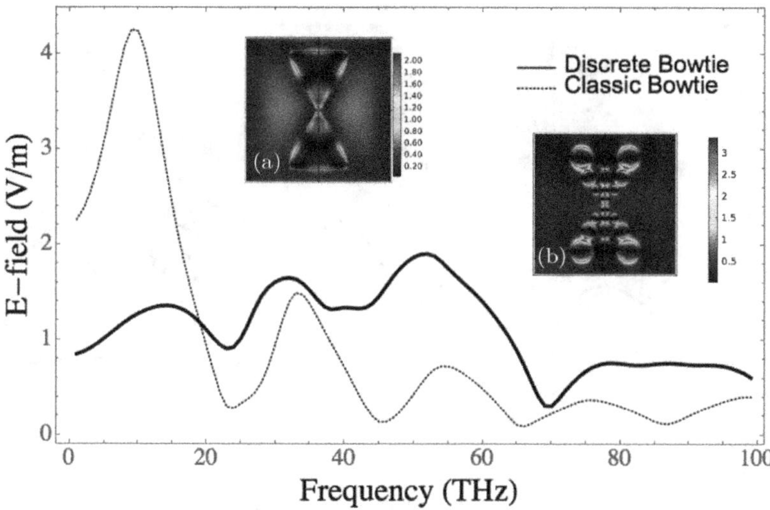

Fig. 12.5. Magnitude of the electric field at the gap of the classical bowtie nanoantenna and discrete bowtie nanoantenna as a function of frequency. Electric field pattern for the (a) classic bowtie antenna and (b) discrete bowtie antenna at a frequency of 28.3 THz (10.6 μm. in wavelength).

simulations, it can be seen how the discrete antenna shows a larger number of hot spots than the classical bowtie antenna and that the bandwidth, considered when the E-field at the gap decreases to half its maximum value, goes from approximately 1 to 20 THz for the classical bowtie antenna and from 5 to 65 THz in the discrete bowtie antenna centered at 35 THz.

Figure 12.6 shows the magnitude of the E-field at the gap of a discrete bowtie antenna and for two fractal bowtie antennas, one with one fractal iteration (Fractal Bowtie 1) and the other with two fractal iterations (Fractal Bowtie 2). The discrete bowtie antenna also has a higher number of electric-field hot spots than the fractal antennas and also a larger bandwidth. The fractal antennas have a bandwidth of 1–15 THz and 27–37 THz for the Fractal 1 bowtie antenna and 1–12.5 THz and 25–30 THz for the Fractal 2 bowtie antenna.

Figures 12.7 and 12.8 show the magnitude of the E-field at the gap of the nanoantennas and the E-field pattern at the plane

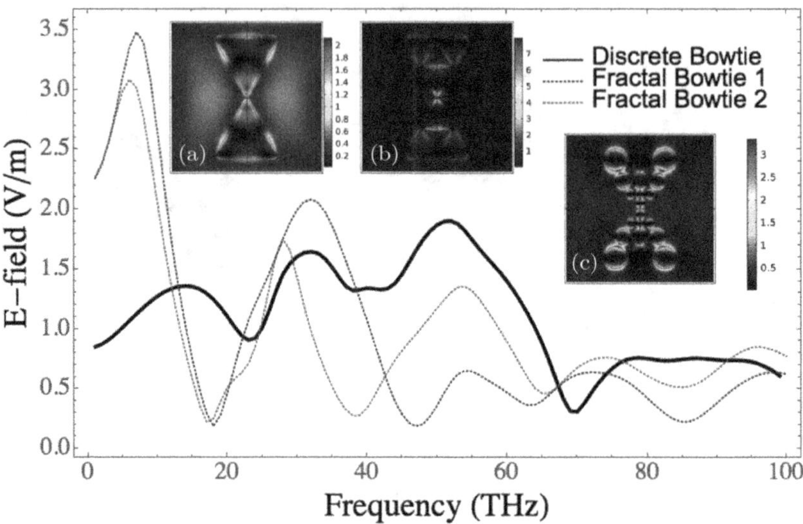

Fig. 12.6.   Magnitude of the electric field at the gap of two different fractal bowtie nanoantennas and a discrete bowtie nanoantenna as a function of frequency.

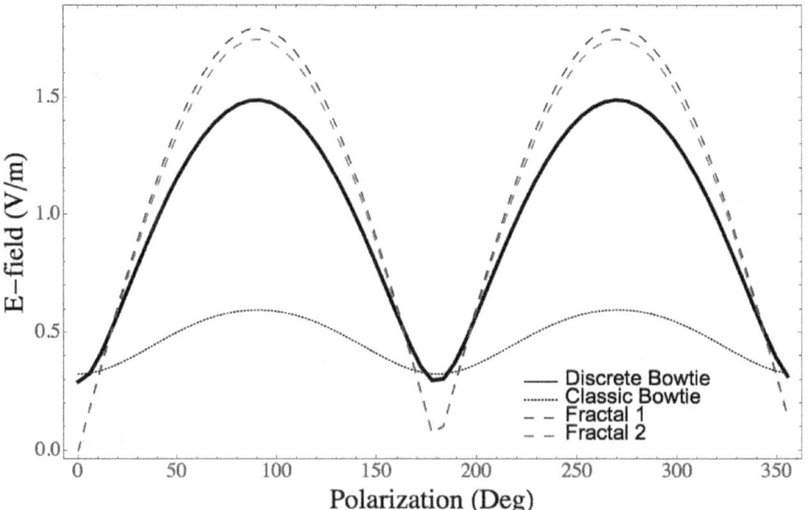

Fig. 12.7.   Magnitude of the electric field at the gap of different bowtie nanoantennas as a function of the angle of polarization for an incident plane wave at 28.3 THz.

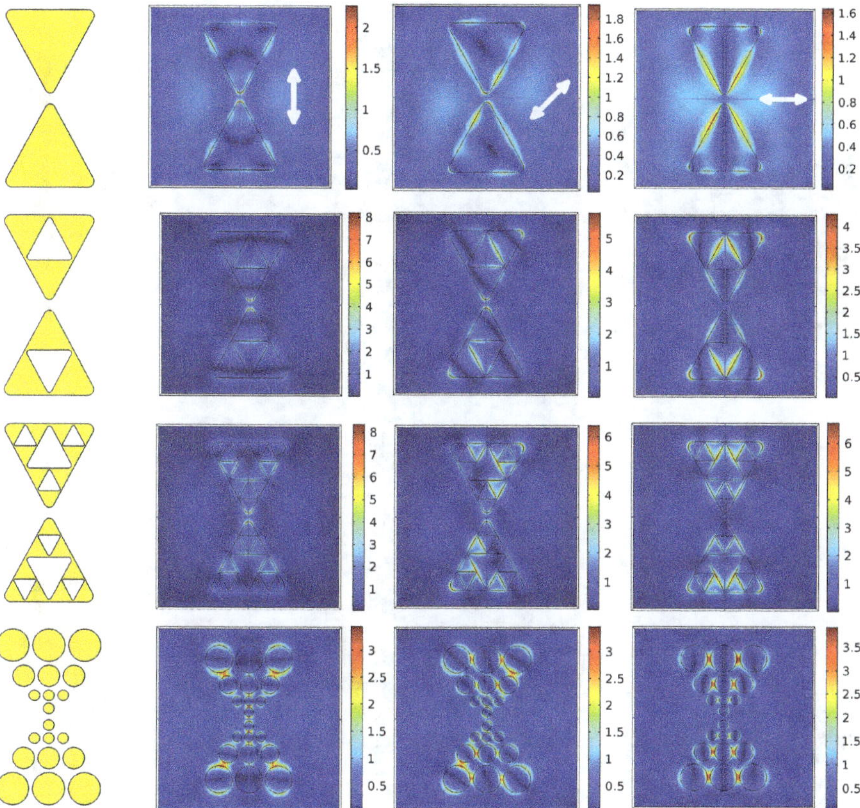

Fig. 12.8. E-field pattern at the nanoantenna plane for different types of bowtie nanoantennas for polarization parallel to the main antenna axis, 45° polarization and perpendicular polarization to the main antenna axis at a fixed frequency of 28.3 THz (10.6 μm. in wavelength).

of the nanoantennas, respectively, for classical, discrete and fractal nanoantennas as a function of polarization. From Fig. 12.7, it can be seen that the polarization sensitivity is higher for discrete and fractal antennas compared to their classical counterparts. This is only measuring the E-field at the gap of the antenna. If we take into account the whole antenna plane (Fig. 12.8), it can be seen how different polarizations still generate E-field hot spots at different locations in the antenna. This is especially noticeable in the discrete

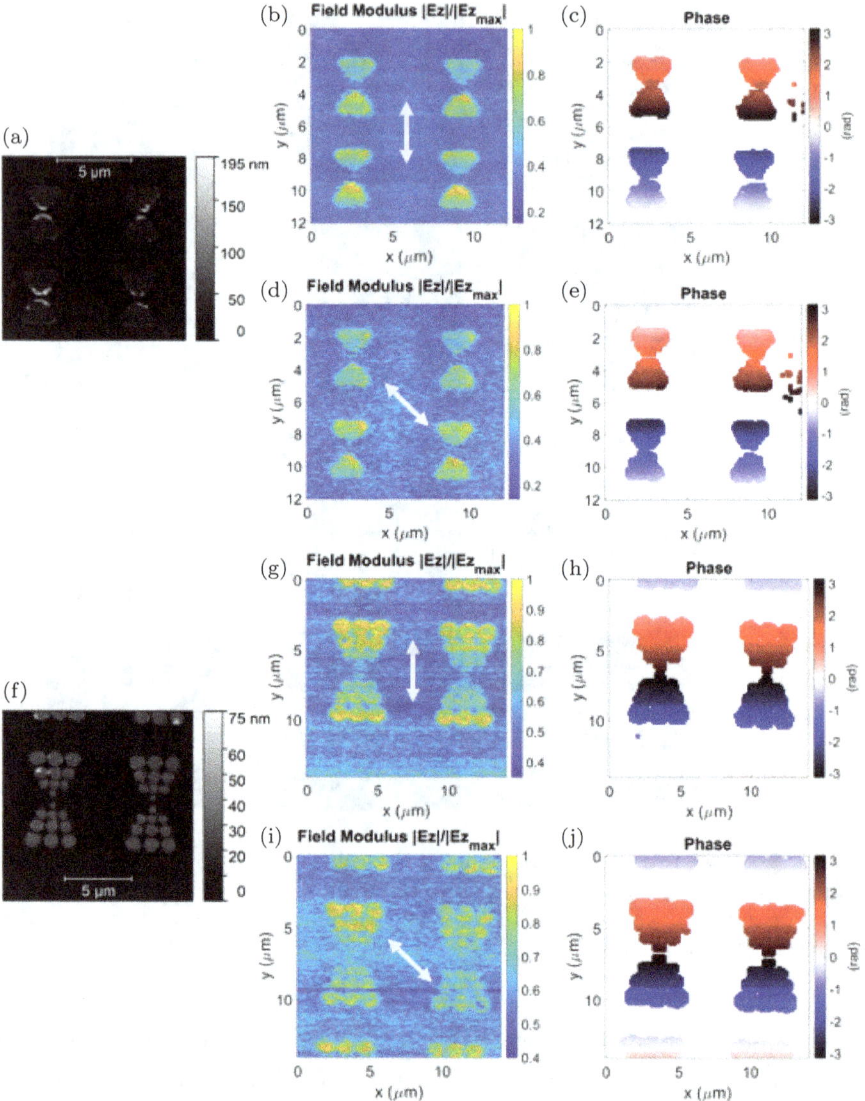

Fig. 12.9. Experimental near-field s-SNOM measurements. AFM topography of the classic bowtie and discrete bowtie are shown in (a) and (f), respectively. Modulus of the near field is shown in (b), (d), (g) and (i), and phase is shown in (c), (e), (h) and (j). Scans (b) and (c) correspond to the classical bowtie when illuminated with vertically polarized light, while (d) and (e) when illuminated with linearly polarized light at $-45°$. Scans (g) and (h) correspond to the discrete bowtie when illuminated with vertically polarized light, while (i) and (j) when illuminated with linearly polarized light at $-45°$.

bowtie nanoantenna, where the largest amount of hot spots are present for different polarizations.

Figure 12.9 shows the magnitude and phase of the near field of the classical and discrete bowties. For the classical bowtie, under vertically polarized illumination, the hot spots are greater in the gap and in the edges, in agreement with the simulations. When illuminated with linearly polarized light at $-45°$, hot spots are predominantly localized to the right side of the antenna. In both cases, there is a phase change of π across the bowtie in the vertical direction. On the other hand, the discrete bowties show hot spots over a greater region of the antenna than in the classical counterpart, with higher concentrations toward the upper and lower ends when they are illuminated with vertically polarized light. When illuminated with linearly polarized light at $-45°$, the near-field distribution is no longer symmetric with respect to the horizontal. Results suggest that the electric field is greater toward the ends of the bowtie and tends to be stronger in the lower-right edges of the circular elements that compose the bowtie. This behavior agrees with the simulations shown in Figure 12.8. For both polarization orientations, the phase difference between the edges of the bowties is π.

Several antenna parameters can be enhanced by discretizing the shape of the antenna. In the case of discrete bowtie antennas, numerical simulations and near-field measurements show that the antenna bandwidth and the number of hot spots can be increased by discretizing the bowtie antenna. This enhancement of antenna properties can be of interest for applications where an increase in electric field not only in a single spot but in several locations in the same structure is desired. In the particular case of discrete antennas, those electric field hot spots appear over a broad range of frequencies, making these antennas of interest for several applications, such as surface-enhanced Raman spectroscopy, biosensing and plasmon generation.

Chapter 13

Plasmonic Biosensors

Point-of-care (POC) testing consists of performing a diagnostic test outside of a laboratory which will produce a rapid and reliable result. POC testing allows for the screening and treatment process to be completed during a single encounter, improving access to care and patient outcome. Studies comparing POC with laboratory testing in an emergency room (ER) setting have demonstrated that POC testing aided in the triage of patients reduces care time.

Advances in several fields, particularly the field of microelectromechanical systems (MEMS) have impacted the development of low-cost biosensors designed for real-time identification of biomarkers. Plasmonic materials can also be used for POC sensors. Surface plasmon resonances are very sensitive to changes in the optical properties of media surrounding the metallic nanostructures, making them a good alternative for the detection of disease markers.

Figure 13.1 shows a diagram of a biomedical device with a plasmonic-enabled detection scheme where several arrays of nanoholes are used as plasmonic biosensing elements. In this case, each nanohole array would be used to detect a specific disease marker. Nanohole arrays support enhanced light transmission at wavelengths that sustain surface plasmon excitations (Brolo, 2012).

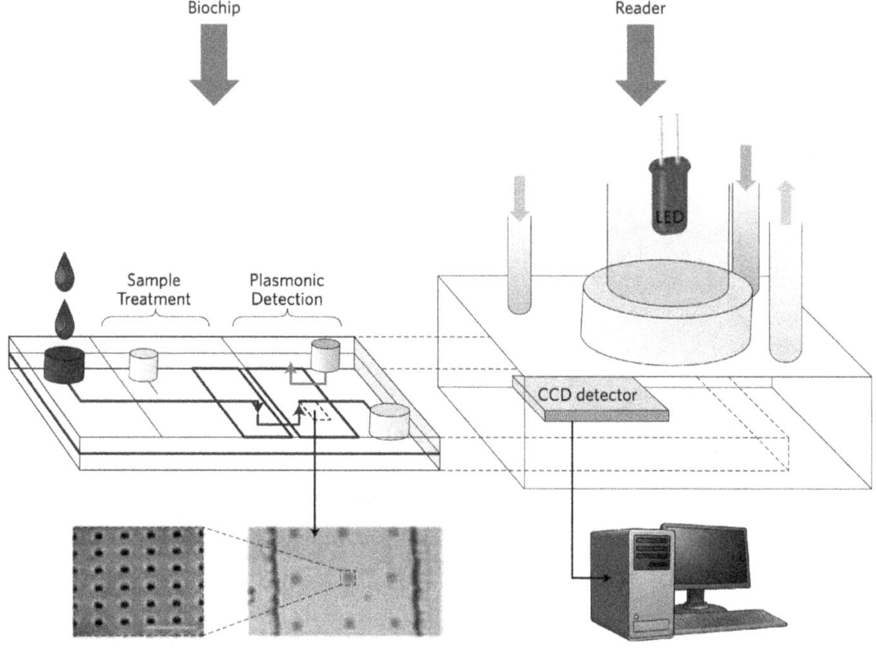

Fig. 13.1. Diagram of a plasmonic biosensor. Taken from (Brolo, 2012).

The plasmonic sensor can be implemented with a low-cost light-emitting diode (LED) and a charge-coupled device (CCD) detector. The small footprint of the plasmonic sensor array allows for the biochemical characterization of a sample from a very small volume ($< 10\mu$l) of the initial biological material.

Surface plasmons exist at metallic nanostructures when conduction electrons oscillate in resonance with an external optical field. These oscillations result in subwavelength-light confinement and electric-field enhancement near the metallic surface. The electric-field enhancement can be used to increase the signal obtained from Raman or fluorescence measurements, resulting in surface-enhanced Raman spectroscopy or surface-enhanced fluorescence; these two techniques can also be used to detect disease markers.

13.1 Plasmonic-Based Refractometric Sensors

A popular configuration for a refractometric sensor involves generating surface plasmons through a prism coupling, either using the Otto or the Kretschmann configurations. Figure 13.2 shows a plasmonic-based refractometric sensor made of a thin metal film on a prism.

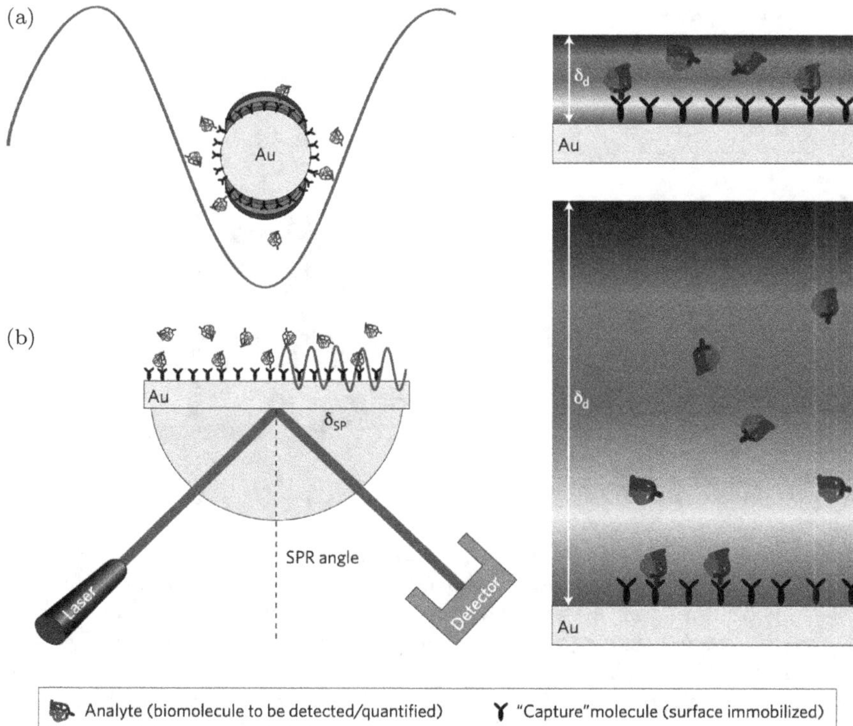

Fig. 13.2. Schematic representation of localized surface plasmon resonances: (a) using a metallic particle much smaller than the wavelength; (b) using a metal–dielectric system in the Kretschmann configuration. The color gradient in each magnified illustration represents the field intensity distribution (the enhanced field at the surface (red) decreases toward the dielectric medium (blue). Taken from (Brolo, 2012).

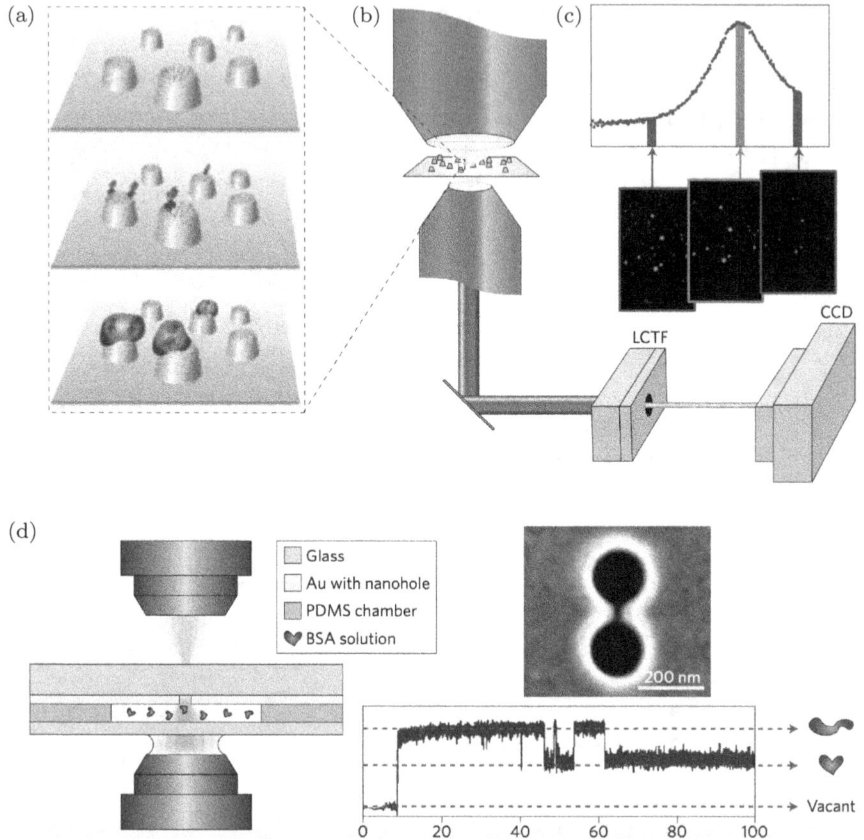

Fig. 13.3. Schematic representation of two plasmonic single-molecule detection schemes: (a) gold nanoparticles dispersed in a glass substrate capturing a single molecule of the analyte of interest, which changes the refractive index around the nanoparticles; (b) several nanoparticles simultaneously interrogated using dark-field imaging; (c) image processing allows the monitoring of the scattering spectrum of each nanoparticle (LCTF: liquid-crystal tunable filter); (d) the intensity of monochromatic light transmitted through a single aperture is modulated by the presence of a single molecule (BSA: bovine serum albumin). Taken from (Brolo, 2012).

The metallic film is functionalized, which means it has been chemically modified to bind to certain biomolecules with a high degree of selectivity. The electromagnetic field generated by the surface plasmons decays exponentially with distance from the surface, with a typical decay length of δ_d.

For propagating surface plasmons, δ_d is in the order of half the resonance of the wavelength (a few hundred nanometers for visible light), while for localized surface plasmons, δ_d is significantly smaller (around 20 nm for a 30 nm diameter spherical gold nanoparticles) (Brolo, 2012).

Surface plasmon resonances are very sensitive to variations in the permittivity of the surrounding media. The biomolecules that are bound by the functionalized metallic film in the refractometric sensor change the effective permittivity of the surrounding media, resulting in a very sensitive sensing mechanism.

13.2 Single-Molecule Sensing with Plasmonic Biosensors

The high sensitivity of localized surface plasmon resonances can be used to monitor single-molecule chemical events. These events can be detected by using nanoparticles or nano-apertures in metal films, which also allow the trapping and manipulation of single particles (Figure 13.3(a)) and single molecules (Figure 13.3(d)).

Bibliography

Aizpurua, J. and R. Esteban. Optical antennas for field-enhanced spectroscopy. In M. Agio and A. Alù (eds.) *Optical Antennas* (pp. 64–80). Cambridge: Cambridge University Press (2013).

Asiala, S.M. and Z.D. Schultz. Characterization of hotspots in a highly enhancing SERS substrate. *Analyst (Lond.)* 136(21), 4472–4479 (2011).

Atwater, H. and A. Polman. Plasmonics for improved photovoltaic devices. *Nature Materials*, 9, 205213 (2010).

Baffou, G. and R. Quidant. Thermo-plasmonics: Using metallic nanostructures as nano-sources of heat. *Laser & Photonics Reviews*, 7, 171–187 (2013).

Balanis, C.A. *Antenna Theory*. John Wiley & Sons (1997).

Barnes, W., A. Dereux, and T. Ebbesen. Surface plasmon subwavelength optics. *Nature*, 424, 824–830 (2003).

Boswell, A.G.P. Guglielmo Marconi's antenna design. *Tenth International Conference on Antennas and Propagation (ICAP)* (1997).

Brolo, A. Plasmonics for future biosensors. *Nature Photon*, 6, 709–713 (2012).

Caiazzo, M., S. Maci, and N. Engheta. A metamaterial surface for compact cavity resonators. *IEEE Antennas and Wireless Propagation Letters*, 3, 261 (2004).

Cakmakyapan, S., N.A. Cinel, A.O, Cakmak, and E. Ozbay. Validation of electromagnetic field enhancement in near-infrared through Sierpinski fractal nanoantennas. *Optics Express*, 22, 19504–19512 (2014).

Camacho, J.M. and A.I. Oliva. Morphology and electrical resistivity of metallic nanostructures. *Microelectronics Journal*, 36(36), 555–558 (2005).

Campbell, S.A. *The Science and Engineering of Microelectronic Fabrication*. Oxford University Press (1996).

Chen, J., D. Wang, J. Xi, L. Au, A. Siekkinen, A. Warsen, Z.-Y. Li, H. Zhang, Y. Xia, and X. Li. Immuno gold nanocages with tailored optical properties for targeted photothermal destruction of cancer cells. *Nano Letters*, 7(5), 1318–1322 (2007).

Chong, N. and H. Ahmed. Antenna-coupled polycrystalline silicon air-bridge thermal detector for mid-infrared radiation. *Appl. Phys. Lett.*, 71, 1607–1609 (1997).

Codreanu, I., C. Fumeaux, D.F. Spencer, and G.D. Boreman. Microstrip antenna-coupled infrared detector. *Electronics Letters*, 35, 2166–2167 (1999).

Codreanu, I., F.J. Gonzalez, and G.D. Boreman. Detection mechanisms in microstrip dipole antenna-coupled infrared detectors. *Infrared Physics & Technology*, In Press (2003).

Derkacs, D., S.H. Lim, P. Matheu, W. Mar, and E.T. Yu. Improved performance of amorphous silicon solar cells via scattering from surface plasmon polaritons in nearby metallic nanoparticles. *Applied Physics Letters*, 89(9), 093103 (28 August 2006).

Dong, L., X. Yang, C. Zhang, B. Cerjan, L. Zhou, M.L. Tseng, Y. Zhang, A. Alabastri, P. Nordlander, and N.J. Halas. Nanogapped Au antennas for ultrasensitive surface-enhanced infrared absorption spectroscopy. *Nano Letters*, 17(9), 5768–5774 (2017).

Elwenspoek, M. and R. Wiegerink. *Mechanical Microsensors*. Springer (2001).

Faris, S.M., B. Fan, and T.K. Gustafson. Electron tunneling currents at optical frequencies. *Applied Physics Letters*, 27(11), 629–631 (1975).

Fatermans, J., A.J. den Dekker, K. Muller-Caspary, I. Lobato, C.M. O'Leary, P.D. Nellist, and S. Van Aert. Single atom detection from low contrast-to-noise ratio electron microscopy images. *Physical Review Letters*, 121, 056101 (2018).

Feynman, R.P. There's plenty of room at the bottom: An invitation to open up a new field of physics. *Caltech Engineering and Science*, 23(5), 22–36 (1960).

Fumeaux, C., W. Herrmann, F.K. Kneubuhl, and H. Rothuizen. Nanometer thin-film Ni-NiO-Ni diodes for detection and mixing of 30 THz radiation. *Infrared Physics & Technology*, 39, 123–183 (1998).

Gall, D. Electron mean free path in elemental metals. *The Journal of Applied Physics*, 119(8), 085101 (2016).

Ghorbani, F., S. Beyraghi, J. Shabanpour, *et al.* Deep neural network-based automatic metasurface design with a wide frequency range. *Scientific Reports*, 11, 7102 (2021).

Giannini, V., A.I. Fernández-Domínguez, S.C. Heck, and S.A. Maier (2011). Plasmonic Nanoantennas: Fundamentals and Their Use in Controlling the Radiative Properties of Nanoemitters. *Chemical Reviews*, 111(6), 3888–3912 (https://pubs.acs.org/doi/10.1021/cr1002672).

Govorov, A.O. and H.H. Richardson. Generating heat with metal nanoparticles. *Nano Today*, 2(1), 30–38 (2007).

Grossman, E.N., J.E. Sauvageau, and D.G. McDonald. Lithographic spiral antennas at short wavelengths. *Appl. Phys. Lett.*, 59, 3225–3227 (1991).

Guisbiers, G., M. Kazan, O. Van Overschelde, M. Wautelet, and S. Pereira. Mechanical and thermal properties of metallic and semiconductive nanostructures. *The Journal of Physical Chemistry C*, 112(11), 4097–4103 (2008).

Hasnain, G., A. Dienes, and J.R. Whinnery. Dispersion of picosecond pulses in coplanar transmission lines. *IEEE Trans on Microwave Theory and Techniques*, 34(6), 738–741 (1986).

Hatzakis, M., D. Hofer, T.H.P. Chang. New hybrid (e-beam/x-ray) exposure technique for high aspect ratio microstructure fabrication. *J. Vac. Sci. Technol.* 16(6), 1631–1634 (1979).

Herzog, J.B., M.W. Knight, and D. Natelson. Thermoplasmonics: Quantifying plasmonic heating in single nanowires. *Nano Letters*, 14(2), 499–503 (2014).

Holloway, C.L., D.C. Love, E.F. Kuester, A. Salandrino, N. Engheta. Subwavelength resonators: On the use of metafilms to overcome the 1/2 size limit. *IET Microwave Antennas & Propagation*, 2, 120 (2008).

Hsu, K.H., J.H. Back, K.H. Fung, P.M. Ferreira, M. Shim, and N.X. Fang. SERS EM field enhancement study through fast Raman mapping of Sierpinski carpet arrays. *Journal of Raman Spectroscopy*, 41(10), 1124–1130 (2010).

Hu, J., S. Bandyopadhyay, Y.-h. Liu, and L.-y. Shao. A review on metasurface: From principle to smart metadevices. *Frontiers in Physics*, 8, 586087 (2021).

Hwang, T.L., S.E. Schwarz, and D.B. Rutledge. Microbolometers for infrared detection. *Applied Physics Letters*, 34(11), 773–776 (1979).

Ingold, G.-L., A. Lambrecht, and S. Reynaud. Quantum dissipative Brownian motion and the Casimir effect. *Physical Review E*, 80(4), 041113 (2009).

Jalalabadi, T., M. Drewery, P. Tremain, J. Wilkinson, B. Moghtaderi, and J. Allen. The impact of carbonate salts on char and gas evolution during slow pyrolysis of biomass, cellulose, and lignin. *Sustainable Energy Fuels*, 4, pp. 5987–6003 (2020).

Jang, Y.H., Y.J. Jang, S. Kim, L.N. Quan, K. Chung, and D.H. Kim. Plasmonic solar cells: From rational design to mechanism overview. *Chemical Reviews*, 116(24), 14982–15034 (2016).

Jauffred, L., A. Samadi, H. Klingberg, P.M. Bendix, and L.B. Oddershede. Plasmonic heating of nanostructures. *Chemical Reviews*, 119(13), 8087–8130 (2019).

Karamehmedovic, M., R. Schuh, V. Schmidt, T. Wriedt, C. Matyssek, W. Hergert, A. Stalmashonak, G. Seifert, and O. Stranik. Comparison of numerical methods in near-field computation for metallic nanoparticles. *Optics Express*, 19, 8939–8953 (2011).

Kik, P.G. and M.L. Brongersma. *Surface Plasmon Nanophotonics*. In Brongersma, M.L., Kik, P.G. (eds.). Springer Series in Optical Sciences, Vol. 131. Springer, Dordrecht (2007).

Kildishev, A.V., A. Boltasseva, and V.M. Shalaev. Planar photonics with metasurfaces. *Science*, 339, 1232009 (2013).

Klimov, V. *Nanoplasmonics* (1st ed.). Pan Stanford Publishing (2014).

Klinkova, A., M. Rachelle, M. Choueiri, and E. Kumacheva. Self-assembled plasmonic nanostructures. *Chemical Society Reviews*, 43, 3976–3991 (2014).

Kogure, T. Chapter 2.9 — Electron Microscopy (eds), F. Bergaya, G. Lagaly, *Developments in Clay Science*, Elsevier, Vol. 5, pp. 275–317 (2013).

Lacy, F. Developing a theoretical relationship between electrical resistivity, temperature, and film thickness for conductors. *Nanoscale Research Letters*, 6, 636 (2011).

Li, Yongqian. *Plasmonic Optics: Theory and Applications*. SPIE Press (2017). (h ttps://spie.org/Publications/Book/2263756).

Liu, S-Y., L. Huang, J-F. Li, C. Wang, Q. Li, H-X. Xu, H-L. Guo, Z-M. Meng, Z. Shi, and Z-Y. Li. Simultaneous excitation and emission enhancement of fluorescence assisted by double plasmon modes of gold nanorods. *The Journal of Physical Chemistry C*, 117(20), 10636–10642 (2013).

Londono-Calderon, A., D. Bahena, and M.J. Yacaman. Controlled synthesis of Au@AgAu YolkShell Cuboctahedra with well-defined facets. *Langmuir*, 32(30), pp. 7572–7581, (2016).

Lyshevski, S.E. *MEMS and NEMS: Systems, Devices and Structures.* CRC Press (2002).

Matarrese, L.M. and K.M. Evenson. Improved coupling to infrared whisker diode by use of antenna theory. *Applied Physics Letters*, 17, 8–10 (1970).

Mauser, K., S. Kim, S. Mitrovic, D. Fleischman, R. Pala, K.C. Schwab, and H.A. Atwater. Resonant thermoelectric nanophotonics. *Nature Nanotechnology*, 12, 770–775 (2017).

Miwa, K., H. Ebihara, X. Fang, and W. Kubo. Photo-thermoelectric conversion of plasmonic nanohole array. *Applied Sciences*, 10, 2681 (2020).

Moore, G.E. Cramming more components onto integrated circuits. *Electronics*, 114–117 (1965).

Moreno, C., J. Alda, E. Kinzel, and G. Boreman. Phase imaging and detection in pseudo-heterodyne scattering scanning near-field optical microscopy measurements. *Applied Optics*, 56, 1037–1045 (2017).

Moreno, C., J. Méndez-Lozoya, G. González, F.J. González, and G. Boreman. Near-field analysis of discrete bowtie plasmonic nanoantennas. *Microw Opt. Technol. Lett.*, 62, 943–948 (2020).

Motchenbacher, C.D. and F.C. Fitchen. *Low-Noise Electronic Design.* John Wiley & Sons (1973).

Murty, B.S., P. Shankar, B. Raj, B.B. Rath, and J. Murday. Tools to characterize nanomaterials. In *Textbook of Nanoscience and Nanotechnology.* Springer, Berlin, Heidelberg (2013).

Myroshnychenko, V., J. Rodríguez-Fernández, I. Pastoriza-Santos, A.M. Funston, C. Novo, P. Mulvaney, L.M. Liz-Marzán, and F.J. García de Abajo (2008). Modelling the optical response of gold nanoparticles, In *Optical Properties of Metallic Nanoparticles.* Springer Series in Materials Science, Vol. 232.

Nayfeh, M. (ed.). Chapter 16 — Nanoeffects in ancient technology and art and in space, in Micro and Nano Technologies. *Fundamentals and Applications of Nano Silicon in Plasmonics and Fullerines*, Elsevier, pp. 497–518 (2018).

Neikirk, D.P. and D.B. Rutledge. Air-bridge microbolometer for far-infrared detection. *Applied Physics Letters* 44(2), 153–155 (1984).

Ocelic, N., A. Huber, and R. Hillenbrand. Pseudoheterodyne detection for background-free near-field spectroscopy. *Applied Physics Letters*, 89, 101–124 (2006).

Paz-Soldan, D., A. Lee, S.M. Thon, M.M. Adachi, H. Dong, P. Maraghechi, M. Yuan, A.J. Labelle, S. Hoogland, K. Liu, E. Kumacheva, and E.H. Sargent. Jointly tuned plasmonic-excitonic photovoltaics using nanoshells, *Nano Letters*, 13 (4), pp. 1502–1508 (2013).

Pendry, J.B. Negative refraction makes a perfect lens. *Physics Review Letters*, 85, 3966 (2000).

Petti, L. *et al.*, A plasmonic nanostructure fabricated by electron beam lithography as a sensitive and highly homogeneous SERS substrate for bio-sensing applications. *Vibrational Spectroscopy*, 82, 22–30 (2016).

Plummer, J.D., M.D. Deal, and P.B. Gricn. *Silicon VLSI Technology: Fundamentals, Practice and Modeling.* Electronics and VLSI. Prentice Hall (2000).

Polycarpou, A.C. Introduction to the finite element method in electromagnetics. Synthesis lectures on computational electromagnetics. *Springer Nature*, 1, 1–126 (2006).

Pozar, D.M., S.D. Targonski, and H.D. Syrigos. Design of millimeter wave microstrip reflectarrays. *IEEE Transactions on Antennas and Propagation*, 45, 287 (1997).

Rai-Choudhury, P. (Editor). *SPIE Handbook of Microlithography, Micromachining and Microfabrication, Volume 1: Microlithography.* SPIE Press (June 1997).

Ramírez-Elías, M.G. and F.J. González. *Raman Spectroscopy for In Vivo Medical Diagnosis in Raman Spectroscopy.* London, United Kingdom: IntechOpen (2018).

Rogalski, A. *Infrared Detectors.* Gordon and Breach Science Publishers (2000).

Rutledge D.B., D.P. Neikirk, and D.P. Kasilingam. Integrated circuit antennas. In Kenneth J. Button, editor, Millimeter Components and Techniques, Part II, volume 10 of Infrared and Millimeter Waves, chapter 1, pages 1–90. Academic Press (1983).

Ryan, C.G.M. *et al.* A wideband transmitarray using dual-resonant double square rings. *IEEE Transactions on Antennas and Propagation*, 58, 1486–1493 (2010).

Schatz, G.C., M.A. Young, and R.P. Van Duyne. Electromagnetic mechanism of SERS. In: Kneipp, K., Moskovits, M., and Kneipp, H. (eds.) *Surface-Enhanced Raman Scattering. Topics in Applied Physics*, Vol. 103. Springer, Berlin, Heidelberg (2006).

Schmotz, M., J. Maier, E. Scheer, and P. Leiderer. A thermal diode using phonon rectification. *New Journal of Physics*, 13, 113027 (2011).

Schurig, D., J.J. Mock, B.J. Justice, S.A. Cummer, J.B. Pendry, A.F. Starr, and D.R. Smith. Metamaterial electromagnetic cloak at microwave frequencies. *Science*, 314(5801), 977–980 (2006).

Shalf, J. and R. Leland. *Computing Beyond Moore's Law, Computer*, 48(12), 14–23 (2015).

Sheats, J.R. and B.W. Smith (eds.). *Microlithography Science and Technology*, 1st Edition. Marcel Dekker Inc, New York (1998).

Sievenpiper, D., L. Zhang, R.F.J. Broas, N.G. Alexopolous, E. Yablonovitch. High-impedance electromagnetic surfaces with a forbidden frequency band. *IEEE Transactions on Microwave Theory and Techniques*, 47, 2059–2074 (1999).

Tang, C.Y. and Z. Yang. Chapter 8 — Transmission Electron Microscopy (TEM) (eds.), N. Hilal, A.F. Ismail, T. Matsuura, and D. Oatley-Radcliffe, *Membrane Characterization*, Elsevier, pp. 145–159 (2017).

The Nobel Prize in Physics 1965. NobelPrize.org. Nobel Media AB 2021. Sun. 25 Apr 2021. https://www.nobelprize.org/prizes/physics/1965/summary/.

Trügler, A. (2016). Modeling the Optical Response of Metallic Nanoparticles. In *Optical Properties of Metallic Nanoparticles*. Springer Series in Materials Science, Vol. 232.

Tsang, D.-W. and S.E. Schwarz. Detection of 101m radiation with point-contact schottky diodes. *Applied Physics Letters*, 30(6), 263–265 (1977).

Wang, H.L., E.M. You, R. Panneerselvam, *et al.* Advances of surface-enhanced Raman and IR spectroscopies: From nano/microstructures to macro-optical design. *Light: Science & Applications*, 10(1), 161 (2021).

Werner, D.H. and S. Ganguly. An overview of fractal antenna engineering research. *IEEE Antennas Propagation*, 45(1), 38–57 (2003).

Wiley, B.J., S.H. Im, Z-Y. Li, J. McLellan, A. Siekkinen, and Y. Xia. Maneuvering the surface plasmon resonance of silver nanostructures through shape-controlled synthesis. *The Journal of Physical Chemistry B*, 110(32), 15666–15675 (2006).

Wilke, I., W. Herrmann, and F.K. Kneubühl. Integrated nanostrip dipole antennas for coherent 30 THz infrared radiation. *Appl. Phys. B*, 58, 87–95 (1994).

Yang, Y., H.T. Dai, and X.W. Sun. Fractal diabolo antenna for enhancing and confining the optical magnetic field. *AIP Advances*, 4, 017123 (2014).

Yasuoka, Y., T. Shimizu, and K. Gamo. Fabrication of slot antenna coupled warm carrier detectors for millimeters wave radiation. *Microelectronic Engineering*, 445–448 (1998).

Yu, H., Y. Peng, Y. Yang, *et al.* Plasmon-enhanced lightmatter interactions and applications. *npj Computational Materials*, 5(1), 45 (2019).

Zuloaga, J., E. Prodan, and P. Nordlander. Quantum description of the plasmon resonances of a nanoparticle dimer. *Nano Letters*, 9, 887–891 (2009).

Index